T0291008

Sustainable Agriculture

Ever-increasing population growth, combined with ongoing climate change signals that agriculture will face great challenges in ensuring global food security by 2050. Additionally, climate change-driven variations in mean sea level, wave conditions, storm surge, droughts, and river flows could have serious effects on agriculture and other sectors. Considering these factors and the extremely high value and necessity of agriculture worldwide, effective adaptation measures underpinned by reliable climate change impact assessments are essential to conserve soil and water resources and ensure food security. *Sustainable Agriculture: Adaptation Strategies to Address Climate Change by 2050* provides a thorough examination of these issues, and presents in-depth analysis, practical case studies, and numerous examples of adaptation options throughout for various regions of the world.

Features:

- Presents up-to-date, scientifically robust information on climate change projections in Europe, Asia, America, Africa, and Australia.
- Provides pathways to sustainable agricultural options rather than just defining the climate change issue.
- Includes case studies and practical examples throughout the world.
- Presents a framework by which policymakers can begin implementing strategies for improving agricultural productivity.

Sustainable Agriculture
Adaptation Strategies to Address Climate Change by 2050

Zied Haj-Amor, Dong-Gill Kim,
and Salem Bouri

CRC Press
Taylor & Francis Group
Boca Raton London New York

CRC Press is an imprint of the
Taylor & Francis Group, an **informa** business

Designed cover image: Shutterstock

First edition published 2024
by CRC Press
2385 NW Executive Center Drive, Suite 320, Boca Raton FL 33431

and by CRC Press
4 Park Square, Milton Park, Abingdon, Oxon, OX14 4RN

CRC Press is an imprint of Taylor & Francis Group, LLC

ISBN: 978-1-032-51845-9 (hbk)
ISBN: 978-1-032-51846-6 (pbk)
ISBN: 978-1-003-40419-4 (ebk)

DOI: 10.1201/9781003404194

Typeset in Times
by KnowledgeWorks Global Ltd.

Contents

PART I Climate Change by 2050

PART II   Climate Change Effects on Agriculture by 2050

PART III   Adaptation Strategies to Address Climate Change for Global Food Security

Preface

Is our agriculture ready for climate change and global population growth expected by 2050? The United Nations predicts that global population growth will result in 9.4 to 10.1 billion people in 2050, intensifying the pressure on agriculture to meet growing food demands already affected by the impact of climate change. To ensure food security for 9.4 to 10.1 billion people in 2050, global agricultural production must be enhanced by up to 56%. Achieving this enhancement will not be an easy task due to the manifesting effects of climate change on agriculture.

The scientific community cannot remain silent when extreme climate events are happening in many agricultural areas across the world and people are asking questions about the future of agriculture and food security under climate change. It is time to determine accurate facts about our climate in 2050 and prepare our agriculture for these facts through effective adaptation strategies. Nonspecialists, scientists, farmers, decisionmakers, students, planners, administrations, and other stakeholders in the agriculture sector require climate change projections over the coming few decades and accurate climate information to develop appropriate agricultural strategies for the upcoming years. Given the extremely high value of agriculture worldwide, effective adaptation strategies underpinned by reliable climate change impact assessments are essential to ensure agricultural sustainability and achieve food security in 2050. Moreover, the current strong effects of climate change on agriculture give urgency to addressing agricultural adaptation more coherently.

There is an urgent need to get a clear idea about what future climate will be like over the coming few decades and how decision makers and farmers should cope with it. Through thick description, in-depth analysis, concrete examples, and real case studies from over 100 countries, this book seeks to determine what future climate will be like by 2050 throughout the globe (Europe, Asia, North America, and Africa), what effects of climate on agriculture are predicted for countries/regions, and what policies and sustainable on-farm practices must be implemented in the next few years to achieve synchronicity between climate neutrality (i.e., net-zero greenhouse gas emissions) and sufficient global agricultural production targets by 2050.

While it is not too late to achieve climate neutrality by 2050, there is now only a short time to implement adaptation strategies that will contribute to avoiding the worst impacts of global warming on agricultural production and ensuring food security by 2050. This book has six key features: (1) it communicates up-to-date, scientifically robust information on trends in climate change up to 2050 across countries and regions; (2) it analyzes the potential effects of climate change in 2050 on global agricultural productivity; (3) it provides an in-depth discussion of sustainable agriculture and pathways to achieving it; (4) it includes concrete examples and case studies from over 100 countries; (5) it provides information and estimates for both regions (Europe, Asia, North America, and Africa) and individual countries; and (6) it offers 50 essential recommendations for farmers and policymakers to plan for climate change action in 2050 to improve agricultural productivity.

Overall, the output of this book will mainly contribute to promoting sustainable agriculture and achieving the United Nations' 2050 Agenda for climate neutrality and food security. As there is an urgent need for increased commitment to action from national, regional, and global leaders toward synchrony between climate neutrality and food security by 2050, all findings of this book should be key to motivating farmers, researchers, extension workers, governments, and decision-makers to respond to climate change challenges in the agriculture sector as soon as possible. Without urgent and adequate intervention, we may witness catastrophic effects of climate change on agriculture never seen in modern times.

About the Authors

Dr. Zied Haj-Amor's research interests primarily focus on soil and water management under climate change, sustainable agriculture, and climate change resilience in agroecosystems. He has conducted various research projects on soil salinity in Africa and is currently conducting research at Water, Energy and Environment Laboratory, Sfax University in Tunisia. He collaborates extensively with researchers in Africa and Europe to improve productivity of saline soils under climate change.

Dr. Dong-Gill Kim's research interests primarily focus on carbon sequestration, greenhouse gas mitigation, and climate resilience in agroecosystems. He has conducted various carbon and greenhouse gas research projects in South Korea, the USA, Ireland, and New Zealand and is currently working at Wondo Genet College of Forestry and Natural Resources, Hawassa University in Ethiopia. He collaborates extensively with researchers in Africa, Europe, and North and Latin America to enhance food security and greenhouse gas mitigation in smallholder farming systems. He earned his Ph.D. in Environmental Science from Iowa State University in the United States.

Prof. Salem Bouri is a Professor of Earth Sciences at Faculty of Sciences Sfax, Tunisia. He has published over 100 peer-reviewed papers in reputed international journals and conferences. He has conducted many research projects on shallow and deep groundwaters in Tunisia. The current research interests of Prof. Salem Bouri include water management, climate change, geo-informatics, hydrogeology, and sustainable agriculture.

Acknowledgments

The authors would like to thank colleagues from Sfax University (Tunisia) for their insightful comments on the present book and their valuable suggestions. Dr. Dong-Gill Kim acknowledges support from International Atomic Energy Agency (IAEA) Coordinated Research Project (CRP D1.50.20) 'Developing Climate Smart Agricultural Practices for Mitigation of Greenhouse Gases'.

About the Book

This book provides an overview of climate change projections by 2050 across various regions and countries. It analyzes agricultural production under both current and future climate conditions (projected for 2050). Additionally, it discusses the latest policies and sustainable practices that could be used to achieve synchronicity between climate neutrality (i.e., net-zero greenhouse gas emissions) and global agricultural production targets by 2050. The book aims to promote the adoption of the best sustainable farming practices, tailored to local conditions.

The book has six key features: (1) it communicates up-to-date, scientifically robust information on trends in climate change up to 2050 across countries and regions; (2) it analyzes the potential effects of climate change in 2050 on global agricultural productivity; (3) it provides an in-depth discussion of sustainable agriculture and pathways to achieving it; (4) it includes concrete examples and case studies from over 100 countries; (5) it provides information and estimates for both regions (Europe, Asia, North America, and Africa) and individual countries; and (6) it offers 50 essential recommendations for farmers and policymakers to plan for climate change action in 2050 to improve agricultural productivity.

Acronyms and Abbreviations

AFOLU	Agriculture, Forestry and Other Land Use
AFS	Agroforestry Systems
APHRODITE	Asian Precipitation-Highly-Resolved Observational Data Integration Towards the Evaluation of Water Resources
APSIM	Agricultural Production Systems Simulator
ARMS	Agricultural Resource Management Survey
AWD	Alternative Wetting and Drying
BMPs	Best Management Practices
C	Carbon
CAP	Common Agricultural Policy
CC	Cover Cropping
CDHEs	Compound Drought-Heat Wave Events
CERES	Crop Environment Resource Synthesis
CF	Continuous Flooding
CFP	Common Fisheries Policy
CH$_4$	Methane
CO$_2$	Carbon dioxide
CRED	Centre for Research on the Epidemiology of Disasters
CRPV	Climate-Resilient Potato Variety
CWP	Crop Water Productivity
D	Dry cultivation
DL	Deep Learning
DSSAT	Decision Support System for Agrotechnology Transfer
DT	Drought Tolerant
EcoTransIT World	Ecological Transport Information Tool Worldwide
EEA	European Environment Agency
EGD	European Green Deal
ELA	Europe's Land Area
EPIC	Environmental Policy Integrated Climate
ESOTC	European State of the Climate
ET	Evapotranspiration
EU	European Union
Eurostat	Statistical Office of the European Union
FAO	Food and Agriculture Organization
FASSET	Farm Assessment Tool
FRD	Flood Risk Directive
F2F	Farm to Fork strategy
GCMs	Global Climate Models
GDP	Gross Domestic Product
GEO	Group on Earth Observations
GGCM	Global Gridded Crop Model
GHG	Greenhouse gas

GMSL	Global Mean Sea Level
GPCC	Global Precipitation Climatology Centre
GREET	Greenhouse gases, Regulated Emissions, and Energy-use in the Transportation model
GSL	Growing Season Length
GSP	Global Soil Partnership
ICRAF	World Agroforestry Centre
ICZM	Integrated Coastal Zone Management
IEA	International Energy Agency
IFAD	International Fund for Agricultural Development
IFI	International Financial Institution
IGP	Indo-Gangetic Plains
INVEST	Infrastructure Voluntary Evaluation Sustainability Tool
IPCC	Intergovernmental Panel on Climate Change
LDN	Land Degradation Neutrality
LT-LEDS	Long-Term Low Emissions and Development Strategy
MAE	Mean Absolute Error
MC-L	Mixed Crop-Livestock
ML	Machine learning
MONICA	Model for Nitrogen and Carbon in Agro-ecosystems
MOVES	Motor Vehicle Emission Simulator
MVEI	Motor Vehicle Emission Inventory Model
N	Nitrogen
NH$_3$	Ammonia
N$_2$O	Nitrous oxide
nRMSE	Normalized Root Mean Squared Error
NSE	Nash-Sutcliffe Efficiency
NUE	Nitrogen Use Efficiency
OPV	Open Pollinated Variety
PA	Precision Agriculture
PBM	Process-Based Crop Model
PEI	Prince Edward Island
PET	Potential Evapotranspiration
PLSF	Precision livestock smart farming
PR	Potato–Rice rotation
PU	Paddy-Upland
RCM	Regional Climate Model
RCP	Representative Concentration Pathway
RE	Relative Error
RF	Random Forest
RWH	Rainwater Harvesting
RMSE	Root Mean Squared Error
RUSLE	Revised Universal Soil Loss Equation
SDG	Sustainable Development Goal
SFM	Sustainable Forest Management
SLEMSA	Soil Loss Estimation Model for Southern Africa

SLR	Sea Level Rise
SNI	Synthetic Nitrification Inhibitor
SOC	Soil Organic Carbon
SPEI	Standardized Precipitation-Evapotranspiration Index
SSA	Sub-Saharan Africa
STICS	Multidisciplinary Simulator for Standard Crops
SWAT	Soil and Water Assessment Tool
THP	Texas High Plains
UK	United Kingdom
UN	United Nations
UNDO	United Nations Development Organization
UNEP	United Nations Environment Programme
UNFCCC	United Nations Framework Convention on Climate Change
USA	United States of America
USDA	United States Department of Agriculture
USDA-ARS	Agricultural Research Service of the United States Department of Agriculture
VRT-N	Variable Rate Technology Nitrogen
VTP	Variety Test Platforms
WB	World Bank
WCST	Water Conservation and Saving Technology
WEF	Water-Energy-Food
WFD	Water Framework Directive
WFP	World Food Programme
WHO	World Health Organization
WMO	World Meteorological Organization
WNA	World Nuclear Association
WOFOST	World Food Studies
WUE	Water Use Efficiency

General Introduction

Due to the continuous increase in the concentration of greenhouse gases (GHGs) in the atmosphere as a direct result of anthropogenic activities, global average surface temperature has increased by about 0.1°C per decade over recent decades (Aizebeokhai, 2009; IPCC, 2022). This acceleration in global warming has been associated with changes in precipitation patterns (NOAA, 2022), sea level (UCAR, 2023), and more extreme climate events such as floods, heatwaves, and droughts (Horton et al., 2016). On farms, where climate and agriculture are inextricably linked (Arora, 2019), these changes in climatic conditions have led to stresses in crops such as decreased respiration, photosynthesis, and transpiration, ultimately reducing crop yields and nutritional quality and lowering livestock productivity worldwide (Ye et al., 2013; Arora, 2019). In the absence of effective adaptation measures and if GHG emissions continue at their current rate, global crop yields could decrease by up to 30% by 2050 (Global Commission on Adaptation, 2019).

Furthermore, it is worth noting that the world population is expected to reach 9.4 to 10.1 billion by 2050, which would magnify the pressure on farms to meet growing food demands already affected by the impact of climate change (Arora, 2019). Overall, due to expected climate change and global population growth in 2050, global agricultural production must increase by up to 56% relative to 2010 (van Dijk et al., 2021) to meet food demands for at least 9.4 billion people (United Nations, 2019). Ensuring global food security for 9.4 to 10.1 billion people by 2050 will be a great challenge for the international community and will increase the difficulty of managing food risk.

As such, there is an urgent need to evaluate the potential impacts of climate change on global agricultural production by 2050 under various socio-economic and climate change scenarios and develop the most effective and sustainable agricultural adaptation options that must be implemented over the next few years to avoid food insecurity by 2050 (Teng et al., 2015). Accurate information on climate projections for 2050 is highly needed for this purpose (IPCC, 2022). Moreover, as the international community is committed to limiting the global average surface temperature increase to 1.5°C above pre-industrial levels through cutting GHG emissions to close to zero by 2050 (i.e., achieving climate neutrality), future agricultural activities require a transition from commercial agriculture to more sustainable agro-ecological farming that focuses on implementing environmentally friendly agricultural options that can simultaneously promote agricultural productivity and restrain further agricultural GHG emissions (Roesch-McNally et al., 2017). A successful transition is highly required to meet the needs of stabilizing agricultural production, ensuring global food security, and realizing the climate neutrality target by 2050 (Zhao et al., 2023).

In this context, this book seeks to analyze different policies and various sustainable on-farm practices that could be used together to achieve synchronicity between climate neutrality and sufficient global agricultural production targets by 2050. The book is structured into three parts. Part I focuses on climate change projections for 2050 across regions and countries. Accurate information on climate projections for

2050 and trends in climate change over the coming decades is key for analyzing the likely effects of climate change on global agriculture and for developing appropriate adaptation measures that can improve agricultural productivity under changing climatic conditions. Based on the outcomes of Part I, Part II describes how climate change would affect agricultural production and food security by 2050 throughout the globe. Finally, based on concrete examples and real case studies from over 100 countries, Part III provides an in-depth discussion of critical components of sustainable agriculture such as cover cropping, cultivation of improved crop varieties, agroforestry, improved nitrogen use efficiency, efficient crop rotations, promotion of new climate-tolerant crops, efficient irrigation under droughts, sustainable management of rice fields, organic agriculture, and sustainable forest management. Essential recommendations for better agricultural performance under climate change by 2050 are provided at the end of the book.

Part I

Climate Change by 2050

Accurate 2050 climate information is highly needed to identify appropriate adaptation measures and improve agricultural productivity under climate change issue. It is hoped that the first part of this book will help to provide decision makers, specialists, farmers, students, and nonspecialists with easy access to 2050 climate information for adaptation planning.

1 Global Climate Change
Its Main Indicators by 2050

1.1 INTRODUCTION

According to the World Meteorological Organization (WMO), climate is defined as the weather pattern of a specific area over an extended period, typically at least 30 years. Climate variables such as air temperature, precipitation, relative humidity, and wind speed change over time and are influenced by several factors, including geographic location (latitude, longitude, and elevation), topography, proximity to the sea and ocean currents, prevailing wind direction, and vegetation (Karl et al., 1999). Numerous studies have shown that global, regional, and local climates are changing rapidly (Carter 2011; Aloysius et al., 2016; Bhuyan et al., 2018; Meinshausen et al., 2022). The IPCC (2018) estimates that by 2050, temperatures may rise by 1.5°C above preindustrial levels with potentially severe impacts on natural and human systems if no action is taken to mitigate climate change (Yerlikaya et al., 2020). The term "climate change" often refers to changes in climate resulting from human activities and greenhouse gas (GHG) emissions (IPCC, 2013). It describes statistically significant trends in average climate variables over time for a specific region or the world. These trends are identified by comparing current climate patterns with past or typical patterns. Describing long-term trends in climate variables is crucial for detecting climate change (IPCC, 2013).

Understanding past and recent climate change has garnered global attention, leading to the development of numerous climate models (Abbass et al., 2022). These models provide past and future climate data for specific areas, allowing climatologists to determine whether abnormal climate events (e.g., high temperatures, droughts, floods) are due to climate change or routine climate variation (IPCC, 2013). They also aid in developing adaptation strategies to reduce the impact of climate change on natural and human systems (Estrada et al., 2022). However, due to the complexity of the climate system, future climate projections can be uncertain (Curry and Webster, 2011). To obtain accurate projections, it is essential to identify potential uncertainties and use appropriate methods and tools (e.g., bias correction, drift removal) to manage them (Estrada et al., 2022).

The Coupled Model Intercomparison Project hosts many general climate models and is commonly used to understand future climate responses to different GHG emission scenarios (Taylor et al., 2012; Stocker et al., 2013; Hamed et al., 2022). However, adapting Coupled Model Intercomparison Project (CMIP) models for specific needs (e.g., projecting the climate for a small area) can be challenging. As a result, many scientists have developed their own regional climate models to address these constraints (Estrada et al., 2022). Using these models in conjunction with observational climate data, several international studies have revealed numerous indicators of climate change such as warming, sea level rise, and rainfall extremes (IPCC 2013; IPCC 2018; IPCC 2021). This chapter discusses the evolution of climate modeling and describes indicators of global climate change by 2050. It is structured into three sections: The first section presents the main causes of global climate change. The second section analyzes how climate models are used to study changing climates. The final section discusses aspects of future global climate by 2050.

1.2   MAIN CAUSES OF GLOBAL CLIMATE CHANGE

Climate change is widely regarded as the greatest global issue of the century (Hansen et al., 2000). Climate model simulations and observational data confirm that GHG, particularly carbon dioxide (CO_2), methane (CH_4), and nitrous oxide (N_2O), emitted through human activities and land-use changes such as deforestation are the primary causes of recent climate change (IPCC, 2014). CO_2 was the largest contributor, accounting for around three-quarters (74.4%) of total emissions. CH_4 contributed 17.3%; N_2O 6.2%; and other emissions (HFCs, CFCs, SF6) 2.1% (IPCC, 2014). As concentrations of these gases increase, they accumulate in the atmosphere, raising global air temperatures and causing numerous changes in the atmosphere, on land and in the oceans (IPCC, 2014). Over the past 170 years, CO_2 concentrations have risen by approximately 50% (NASA, 2020). Human activities such as electricity generation, transportation, agriculture, and waste management (Figure 1.1) have increased GHG concentrations in the atmosphere (IPCC, 2014). It is estimated that these activities emit around 50 billion tons of GHGs annually (Ritchie and Roser, 2020).

1.2.1   ELECTRICITY GENERATION

Electricity generation is a major contributor to huge CO_2 emissions (Abdallah and El-Shennawy, 2013). According to the World Nuclear Association (WNA), in 2022, over 40% of global CO_2 emissions resulted from the combustion of fossil fuels (coal, oil, natural gas) for electricity generation. This significant proportion makes the electricity sector the largest contributor to climate change. A continuous increase

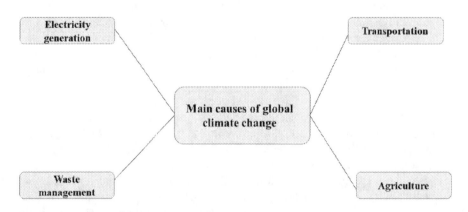

FIGURE 1.1   Main causes of global climate change.

in global electricity consumption at a rate of 8.6% per year (Liu, 2015) may be the primary cause of this high percentage. A study by Cole et al. (2005) found that fossil fuel combustion and associated CO_2 emissions contributed to a 0.3°C increase in global average temperature above preindustrial levels. China and the United States were identified as the largest emitters of CO_2 through electricity generation activities due to their status as the world's largest generators of fossil-fueled electricity (IPCC, 2014). The IPCC (2014) confirmed that CO_2 emissions occur during electricity generation regardless of technology used; however, emissions are more intensive for coal combustion technology and minimal when renewable technologies such as wind are employed. Thus, selection of electricity generation technology plays a crucial role in reducing the environmental footprint of electricity.

1.2.2 TRANSPORTATION ACTIVITIES

Following the electricity sector, transportation is the second largest source of climate change globally and contributes to 20% of global GHG emissions (Albuquerque et al., 2020). The majority of this percentage is attributed to road transportation (ADB, 2010). It is important to note that vehicle operation is not the only source of GHG emissions generated by road transportation; emissions can also result from road construction activities (e.g., excavation, material transport) and maintenance and rehabilitation (Angelopoulou et al., 2009; Albuquerque et al., 2020). Thus, to investigate the impact of the transportation sector on GHG emissions, all these sources must be considered. From 1970 to present day, China, the United States, India, Russia, Japan, Germany, the United Kingdom, and France have been ranked as the world's largest CO_2 emitters via transportation sector (Albuquerque et al., 2020). In recent decades, many statistical models have been developed for predicting GHG emissions from transportation projects (Smit et al., 2007; Ntziachristos et al., 2009; Chiaramonti et al., 2021). Table 1.1 summarizes the most frequently used statistical models by scientists.

1.2.3 AGRICULTURAL ACTIVITIES

Although agriculture is considered the third largest source of global CO_2 emissions (11% of global GHG emissions), following the electricity and transportation sectors, it is ranked as the largest source of CH_4 and N_2O emissions from crop (e.g., rice cultivation) and livestock (e.g., cattle and sheep) production (IPCC, 2014). Compared to preindustrial levels, CH_4 emissions have doubled while N_2O emissions have risen by 20% (IPCC, 2013). CH_4 emissions are primarily caused by decomposition of animal manures while N_2O emissions result mainly from soil nitrification and denitrification processes significantly impacted by land use and land use change (Uchida et al., 2011; Dorich et al., 2020). It is important to note that crop and livestock production are not the only sources of GHG emissions generated by agriculture; emissions can also result from food systems via many pre- and postproduction processes (e.g., fertilizer manufacturing, food processing, packaging, transport, food waste disposal). Recent estimates by IPCC (2022) and FAO (2021) indicate

TABLE 1.1

The Most Commonly Used Models and Tools for Studying Greenhouse Gas Emissions Associated with Road Transportation

Model/Tool	Description	References
The Infrastructure Voluntary Evaluation Sustainability Tool (INVEST)	It is a free web tool designed to assess the sustainability of road projects to ameliorate them based on many criteria.	FHWA (2015)
The Ecological Transport Information Tool Worldwide (EcoTransIT World)	It is used to evaluate the environmental effect of transporting freight by different transport modes.	Borken et al. (2003)
The Motor Vehicle Emission Simulator (MOVES)	It is mainly applied to determine emissions from cars and trucks under different conditions (e.g., road type).	US EPA (2016)
The Motor Vehicle Emission Inventory model (MVEI)	The model is applied to assess pollutants released by the road projects.	El-Fadel and Bou-Zeid (1999)
COPERT model	It is applied by the majority of European countries to estimate pollution emissions from road transport.	Gkatzoflias et al. (2007)
The greenhouse gases, regulated emissions, and energy use in the transportation model (GREET)	It is used to estimated CO_2 emissions from a vehicle/fuel system.	Wang et al. (2007)

that approximately 17 $GtCO_{2eq}$ were emitted in 2018 due to food systems (31% of total GHG emissions). Thus, to estimate the impact of agriculture on GHG emissions, it is necessary to include food systems as a major factor in climate change.

1.2.4 WASTE MANAGEMENT

According to Gautam and Agrawal (2021), waste management may account for 5% of total GHG emissions, with CH_4 and CO_2 being the primary gases generated in this sector. Kristanto and Koven (2019) found that improper waste management can result in significant GHG emissions. The composition of waste varies by country; for example, developing countries primarily produce biodegradable food materials and yard wastes while developed countries produce more paper and cardboard waste (Joseph et al., 2003; IPCC, 2007). GHG emissions can be generated through various waste management options such as controlled landfill disposal, anaerobic digestion of organic solid waste, open combustion, material recovery facilities, and composting (Gautam and Agrawal, 2021). A study conducted by Kristanto and Koven (2019) in Depok City, Indonesia, revealed that 299,602 kg CO_{2eq} day^{-1} was emitted from the management of 600 tons day^{-1} of solid waste through landfill disposal, 70 tons day^{-1} through open combustion, 60 tons day^{-1} through material recovery facilities, 340 tons day^{-1} through anaerobic digestion, and 40 tons day^{-1} through composting. It is important to note that pre- and post-waste management activities such as storage, collection, transfer/transportation recycling, and dumping can also generate GHG emissions. Therefore, each step in the waste management process has the potential to emit GHGs.

1.3   METHODOLOGIES AND TOOLS FOR ACCURATE CLIMATE CHANGE PROJECTION

Scientists use various methodologies and tools to project the world's climate for future years (typically until 2100) based on different scenarios for GHG concentrations known as Representative Concentration Pathway (RCP) scenarios (e.g., Hamed et al., 2022). Regional Climate Models (RCMs) are used to project climate for specific regions while Global Climate Models (GCMs) are used for global projections (e.g., Hamed et al., 2022). Both models may have uncertainties arising from various sources, highlighting the need to quantify and manage them (e.g., Hamed et al., 2022).

1.3.1   GENERAL CLIMATE MODELS

Global Climate Models (GCMs) are the primary tools used by scientists to obtain large-scale climate information for the entire globe over time based on various RCP scenarios developed by the IPCC (e.g., Hamed et al., 2022). These scenarios represent different concentrations of GHGs (e.g., Hamed et al., 2022). GCMs divide the globe into large 3-D grid cells and use mathematical equations to compute what is happening in each cell (IPCC, 2007). This gridded approach allows for the retrieval of climate information from each cell in various aspects (IPCC, 2007). GCMs are constantly being updated with higher spatial resolution, new physical processes, and biogeochemical cycles. Currently, Phase 6 of the CMIP6 GCMs are the latest versions used to investigate past and future changes in Earth's climate (O'Neill et al., 2016). The development of climate projections may consider outputs from all CMIP6 GCMs or a selection based on context and requirements (e.g., Hamed et al., 2022).

1.3.2   REGIONAL CLIMATE MODELS

Due to their large grid resolution (100–500 km), GCMs may have difficulty providing accurate climate information for small areas with complex climates. Additionally, several physical processes that occur at smaller scales, such as those related to clouds, cannot be properly modeled (e.g., Nguyen et al., 2023). To address these limitations, RCMs have been developed to provide detailed climate information for smaller areas of the globe (usually with a 10–50 km grid interval) (e.g., Nguyen et al., 2023). These models are widely used for generating future regional climate projections (Lee and Cha, 2020). RCMs are developed using dynamical and statistical downscaling methods (Wilby et al., 2002). Dynamical downscaling involves nesting a fine-resolution RCM within a GCM and using the GCM's boundary conditions to drive the RCM (Liang et al., 2019). Statistical downscaling involves developing a statistical relationship between observed climate data over a historical period and predicted climate data from a GCM over the same period using statistical analyses such as regression analysis. This relationship is then used to downscale future local climate variables (Ayar et al., 2016). While dynamical downscaling is widely used, it cannot explicitly remove systematic differences (biases) between the global model and observations like statistical methods can (e.g., Nguyen et al., 2023). This can lead to large uncertainties and may require bias correction for certain applications such as estimating future runoff through hydrological models (Zhang et al., 2020). A combined dynamical-statistical downscaling approach is often recommended as it takes advantage of both methods (Han and Wei, 2010). Around 60 RCMs have been developed worldwide to provide scientists with regional-scale climate information (Zhang et al., 2020).

1.3.3 Uncertainty in Climate Projections

Despite significant advancements in climate modeling, uncertainties associated with future climate projections remain a contentious topic within climate science for the 21st century (e.g., Nguyen et al., 2023). These uncertainties encompass a range of factors, including limitations of climate models, future atmospheric concentrations of GHGs (referred to as "scenario uncertainty"), internal variability within the climate system due to natural causes, and potential inaccuracies arising from downscaling methods (Latif, 2011). Given these various sources of uncertainty, it is necessary to approach climate change projections probabilistically; that is, assigning a likelihood to each potential outcome (Giorgi, 2010). Numerous methods have been developed in recent years to analyze and reduce uncertainties associated with climate projections. For instance, internal variability can be assessed through ensemble simulations using the same global model with varying initial conditions (Giorgi, 2010). Scenario uncertainty can be evaluated by comparing results from multiple models run under different emissions scenarios (Lehner et al., 2020). Improving methods for analyzing uncertainties will continue to be an important focus for future climate projections.

1.4 FUTURE GLOBAL CLIMATE BY 2050

According to the 2022 report of World Meteorological Organization, global climate change indicators such as temperature, precipitation, and sea level changes, as well as droughts, heat waves, and floods have had significant impacts on millions of people and resulted in billions of dollars in damages.

1.4.1 Change in Air Temperature

Climatologists have reported changes in global average air temperature over the past decades. Using 1880 as a reference year, the temperature rose by approximately 0.2°C in 1940 and remained relatively stable from 1940 to 1970 (IPCC, 2022). From 1970 to 2020, the temperature increased again at a rate of approximately 0.18°C per decade (IPCC, 2022). Additionally, it was reported that the years from 2015 to 2020 were among the eight warmest on record (IPCC, 2022). The increase in global average air temperature has been primarily attributed to the emission of GHGs, particularly CO_2, CH_4, and N_2O, from fossil fuel combustion and land use changes during

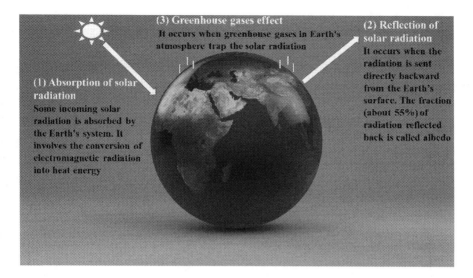

FIGURE 1.2 Greenhouse gases effect.

the period from 1880 to 2020 (IPCC, 2022). These GHGs can persist in the atmosphere for several years and have a high potential to absorb energy, thereby slowing or preventing heat loss to space and contributing to global warming. This phenomenon is commonly referred to as the "Greenhouse effect" (Figure 1.2) (IPCC, 2013).

According to simulations from GCMs, it is projected that the current increase in global average air temperature will continue over the next several decades under various scenarios for GHG concentrations, including very high, medium, and lowest concentrations. The rate of future warming is estimated to be approximately 1.5–2°C by 2050 and 2–4°C by 2100 (IPCC, 2021). The extent of future warming will largely depend on the amount of GHGs emitted into the atmosphere in the coming years (IPCC, 2021).

The United Nations and climatologists have determined that limiting global warming to 1.5°C by 2050 is imperative to prevent a future characterized by frequent disasters such as droughts, wildfires, floods, and storms (IPCC, 2021). Reducing GHG emissions, the primary cause of global warming, is essential to achieving this goal. As called for in the 2016 Paris Agreement, global GHG emissions need to be reduced by 45% by 2030 and reach net zero by 2050. Meeting this 2050 target will require aggressive actions and options, including zero-emission electricity generation, energy efficiency measures, and climate-smart agricultural management practices (Chen et al., 2022).

1.4.2 Change in Precipitation

Precipitation, primarily in the form of rain, fills lakes, rivers, dams, and groundwater reserves that provide water for drinking, irrigation, and industrial uses. However, changing precipitation patterns can limit these uses and have significant impacts on people and the environment. Managing these effects presents a major challenge. Numerous studies (e.g., Karl and Knight, 1998; Wentz et al., 2007) have demonstrated that global precipitation patterns (i.e., amount, distribution, and timing) have changed in recent decades due to global warming. As temperatures rise and air becomes warmer, more water vapor is released from land and oceans into the atmosphere. Trenberth et al. (2003) found that a 1°C increase in global temperature can result in a 7% increase in atmospheric water vapor. This increased water vapor leads to more precipitation events such as intense snowfall and rainfall (Wentz et al., 2007). In fact, very heavy precipitation events have already been observed in many countries (Wentz et al., 2007). It is important to note that this additional precipitation is not evenly distributed across the globe; some regions may actually receive less precipitation than before due to shifts in air and ocean currents caused by global warming (Wentz et al., 2007). The Global Precipitation Climatology Centre (GPCC) reported that dry regions are becoming drier while wet areas are becoming wetter, particularly at mid to high latitudes. This trend is consistent with an overall intensification of the hydrological cycle in response to rising global temperatures. Since 1901, global precipitation has increased at a rate of 1.016 mm per decade (NOAA, 2022). Future precipitation patterns have been simulated using GCMs and RCMs (e.g., Trenberth, 2011; Konapala et al., 2020). These simulations project a significant increase in heavy precipitation intensity globally as global temperatures continue to rise. With an expected increase of 1.5–2°C by 2050, heavy rainfall events are projected to become 1.7 times more likely and 14% more intense (UCAR, 2023). Additionally, it is projected that higher latitude regions will become wetter while regions closer to the equator will become drier (UCAR, 2023). Although GCMs and RCMs have confirmed the increasing trend of intense precipitation events (Wentz et al., 2007; Papalexiou and Montanari, 2019), it is important to note that these models may underestimate increases in short-duration rainfall extremes (Mishra et al., 2012; Papalexiou and Montanari, 2019). Both too little and too much precipitation can pose problems.

1.4.3 Change in Sea Level and Coastal Flooding

According to the NASA records, between 1993 and 2022, the global mean sea level (GMSL) rose by 101 mm at an average rate of 3.4 mm per year. However, it is important to note that some regions have experienced lower or higher rates of sea level rise due to changes in ocean circulation (including changes in ocean temperatures and salinity) or vertical land motion such as local subsidence or uplift (Church, 2019). For instance, along the US coastline, the average sea level rose by 3.8 mm per year from 1997 to 2016 (2017 State of U.S. High Tide Flooding, Appendix 1). Sea level rise is an inevitable consequence of global warming and is driven by two processes: melting of glaciers and ice sheets on land and thermal expansion of ocean water (Vermeer and Rahmstorf, 2009). According to the IPCC Special Report on Ocean and Cryosphere in a Changing Climate (IPCC, 2019), glaciers lost approximately 9,000 billion tons of ice between 1961 and 2016, contributing to a rise in global mean sea level by 27 mm. The same report also indicated that land-based ice in temperate regions such as South America is melting at an accelerated rate.

Precise projections of future sea level rise are difficult to obtain due to large uncertainties in the processes of ice sheet melting and thermal expansion of ocean water (Le Cozannet et al., 2017). However, the application of extra-probabilistic methods (Dubois, 2007; Le Cozannet et al., 2017) can significantly reduce these uncertainties and provide more accurate future sea level rise data using climate models. Based on these models and using 2000 as a reference point, global sea level is projected to continue rising by an additional 15–43 mm by 2050 depending on GHG emission scenarios (e.g., UCAR, 2023). This range represents a rise of 15 mm for the lowest emission scenario and 43 mm for the highest emission scenario (UCAR, 2023). Some low-lying coastal regions may experience even higher rates of sea level rise (UCAR, 2023). In coastal areas, rising sea levels can lead to increased tidal flooding. For example, a sea level rise of 3.8 mm per year from 1997 to 2016 resulted in an average increase in tidal flooding across the United States by 233% (2017 State of U.S. High Tide Flooding, Appendix 1). Sea level rise and its associated tidal flooding have significant impacts on human populations and ecosystems worldwide, particularly in coastal areas where existing infrastructure is not designed to cope with flood events (Doocy et al., 2013). Globally, approximately 250,000 deaths were recorded between 1980 and 2000 due to coastal floods (Nicholls et al., 2007), averaging about 12,000 deaths per year (Shultz et al., 2005). Over the next few decades (2030s–2050s), climate change is expected to increase sea level rise and its associated mortality in several regions worldwide including coastal areas in South Asia, North America, and East Sub-Saharan Africa (World Health Organization, 2014). In light of these risks, there is an urgent need for sea-based defenses in vulnerable coastal areas to improve flood risk management. It is important to note that coastal flood damage and corresponding adaptation strategies are among the most costly impacts of climate change (World Bank, 2010).

1.4.4 DROUGHTS AND HEAT WAVES

Droughts and heat waves are extreme climate events that often occur simultaneously due to global warming and precipitation deficits (Horton et al., 2016). Between 1970 and 2019, droughts caused 650,000 deaths worldwide and resulted in economic losses of US$124 billion globally from 1998 to 2017 (UNCCD, 2022). The frequency of compound drought-heat wave events (CDHEs), also known as hot-dry compounds, has increased over the past two decades in several countries worldwide (Mishra and Singh, 2010; Horton et al., 2016). Globally, using the year 2000 as a reference point, the number and duration of droughts have risen by approximately 30% (UNDRR, 2021). Changes in the frequency of CDHEs have been observed regionally in countries such as China (Ye et al., 2019), the United States (Luo and Wood, 2007), Northwestern Europe (van der Wiel et al., 2022), and Sub-Saharan Africa (Sheffield et al., 2014). These studies confirm that the current and future occurrence of CDHEs in a particular area depends more on precipitation than on warming trends. This can be explained by the fact that local warming will be significant enough that future droughts will always coincide with at least moderately hot extremes even in a world that is 2°C warmer (Bevacqua et al., 2022). A strong increase in the occurrence of CDHEs is predicted under various climate change scenarios (Alfieri et al., 2017; Dosio et al., 2018), but associated prediction uncertainties remain large and poorly understood (see details about these uncertainties in Bevacqua et al., 2022). Therefore, additional efforts are needed to reduce uncertainties associated with future drought predictions. Currently about one-third of the world's population (i.e., 2.3 billion people) are affected by droughts and heat waves, and this number is expected to rise to 75% of the world's population by 2050 if GHG emissions continue to increase without any mitigating actions (UNDRR, 2021). Unlike other aspects of climate change, drought in a particular area is a slow process that occurs over an extended period. This makes it difficult to recognize until its damages have already occurred. As such, accurate real-time prediction of droughts is crucial for global drought preparedness and for mitigating the impacts of CDHEs.

1.5 CONCLUSIONS

In this chapter, we examined the primary indicators of global climate change projected for 2050. The findings indicate that anthropogenic GHG emissions will significantly alter the climate by 2050. Projections include a warmer climate, rising sea levels relative to land, increased frequency and intensity of flood events, and prolonged drought periods. To mitigate these changes, it is imperative to reduce GHG emissions and prepare for the impacts of global warming that are already being observed. Engagement with businesses to decrease their GHG emissions is urgently needed. Implementing measures such as improving energy efficiency, promoting sustainable transportation, and restoring degraded lands can prevent many of the costs and damages associated with climate change by 2050. However, knowledge gaps remain regarding future climate predictions and trends in GHG emissions over the coming decades. Further research is necessary to improve our understanding of GHG emission trends and develop innovative tools to reduce uncertainties in future climate predictions.

EXERCISE (WITH ANSWERS)

Read the above chapter carefully and answer the following five questions:

Question: What is global warming and what are its main cause?

Answer: Global warming has referred to the long-term rise in the average temperature of the Earth's climate system since 1880, when global temperature measurements first began. The primary cause of global warming is the emission of greenhouse gases (GHGs) such as carbon dioxide (CO_2), methane (CH_4), and nitrous oxide (N_2O) from human activities, including burning fossil fuels, transportation, waste management, and agriculture. Climatologists have confirmed that these human activities are the main drivers behind the current global warming trend.

Question: What is the difference between global warming and climate change?

Answer: Global warming refers to the long-term rise in the average temperature of the Earth's climate system. In contrast, climate change encompasses changes not only in temperature but also in other climate parameters such as precipitation, relative humidity, and wind speed over an extended period of time.

Question: How does global warming affect rising sea levels?

Answer: Global warming is a major contributor to rising sea levels through two primary processes. The first process involves the melting of glaciers and ice sheets on land, particularly in Greenland, Antarctica, and mountain glaciers worldwide. The second process is thermal expansion, whereby warmer water occupies more space and causes an increase in ocean volume. Both processes contribute to rising sea levels.

Question: What is a "carbon footprint"?

Answer: A carbon footprint refers to the total amount of GHGs emitted into the atmosphere by an individual or tool system or activity. It is measured in units of carbon dioxide equivalent (CO_{2e}). Precise measurement equipment is required to accurately determine this amount. Scientists have developed various online tools that use standardized methods to calculate the carbon footprint of buildings, cars, boats, airplanes, foods, and industries. Some examples of these online tools include

- Cool Climate Calculator (https://coolclimate.berkeley.edu/calculator)
- Carbon Calculator (https://www.carbonfootprint.com/calculator.aspx)
- UN Carbon Footprint Calculator (https://offset.climateneutralnow.org/footprintcalc)
- TerraPass Carbon Footprint Calculator (https://calculator.terrapass.com/#/)
- EPA Carbon Footprint Calculator (https://www3.epa.gov/carbon-footprint-calculator/)

Question: What is the role of UN to combat climate change?

Answer: The United Nations (UN) plays a key role in facilitating international negotiations on climate change through the United Nations Framework Convention on Climate Change (UNFCCC) and the Intergovernmental Panel on Climate Change (IPCC). The UNFCCC provides a legal framework and principles for international cooperation to stabilize atmospheric concentrations of GHGs and prevent dangerous climate change. The IPCC, composed of climate experts, provides policymakers with accurate climate information, and recommends strategies to reduce GHG emissions and adopt climate change. In 2015, the UN adopted Sustainable Development Goal 13 (SDG 13), "Climate Action," which calls on countries to reduce their GHG emissions and prepare for the impacts of climate change.

2 Climate Change in Europe

Past Trends and Key Indicators by 2050

2.1 INTRODUCTION

Europe, encompassing an area of 10.5 million km^2 and home to approximately 720 million people (UN DESA, 2022), has been a significant contributor to global climate change. Between 1950 and 2012, countries within the European Union (EU) were responsible for approximately 17% of global cumulative GHG emissions (Rocha et al., 2015), making the EU the third-largest emitter after China and the United States. Consequently, studies have confirmed that climate change has impacted all European countries over recent decades (e.g., Thonicke and Cramer, 2006; Gregory et al., 2009; Kovats et al., 2011; Ionita and Nagavciuc, 2021). However, due to variations in climate across Europe, the effects of climate change differ significantly between countries. Many European nations have conducted extensive research on climate change using innovative approaches such as machine learning (ML) and deep learning (DL) models (e.g., Spinoni et al., 2015; Russo et al., 2015; Rajczak et al., 2016; Saloux and Candanedo, 2018; García-León et al., 2021). These studies, along with European State of the Climate (ESOTC) reports published by the European Centre for Medium-Range Weather Forecasts (ECMWF), provide detailed information on key indicators of climate change in Europe including rising temperatures, droughts and heatwaves, extreme weather events, and changing precipitation patterns. These climate change indicators have had profound impacts on ecosystems, economies, agriculture, and human health throughout Europe (EEA, 2012; IPCC, 2019). For instance, the statistical office of the European Union (Eurostat) estimates that climate change has resulted in economic losses of €487 billion for the EU between 1980 and 2020 (Germany: €108 billion; France: €99 billion; Italy: €90 billion). Furthermore, projections from the sixth assessment report of the Intergovernmental Panel on Climate Change (IPCC) suggest that trends in many key climate change indicators will persist across all European countries. In response to these projections, the EU aims to achieve full adaptation to the unavoidable impacts of climate change by 2050 (European Commission, 2020). This will require significant efforts to reduce GHG emissions over the coming decades to meet Europe's target of becoming the world's first carbon-neutral bloc by 2050.

In light of this discussion, this chapter seeks to address several questions: (i) What are the primary drivers of observed climate change in Europe? (ii) What are the key

DOI: 10.1201/9781003404194-3

indicators of climate change observed in Europe over recent decades? (iii) What are potential future climate scenarios for Europe by 2050? (iv) What is Europe's target climate by 2050? Answers to these questions can inform planning and implementation of activities and actions aimed at achieving net-zero GHG emissions by 2050. Data for this chapter were primarily sourced from IPCC reports, ESOTC reports, the European Environment Agency (EEA), the European Commission, the World Meteorological Organization (WMO), NASA climate reports, and numerous journal articles.

2.2 MAIN CAUSES OF CLIMATE CHANGE IN EUROPE

In 2018, total GHG emissions from EU-27 countries and the United Kingdom amounted to approximately 4,400 million tons of carbon dioxide equivalent ($MtCO_{2e}$) (EEA, 2020), making the EU the third largest emitter after China and the USA. Data from the Global Carbon Project (GCP) indicate that GHG emissions in the EU have changed over time as follows: between 1960 and 1979, emissions increased rapidly due to economic growth; between 1980 and 1989, emissions remained relatively stable; and between 1990 and 2019, emissions decreased by 24%, with most of this reduction occurring between 2010 and 2019. This decrease can be attributed to the implementation of various climate and energy policies by EU countries during this period (European Commission, 2020; Scarlat et al., 2022), particularly Germany and the United Kingdom. In fact, these two countries accounted for approximately half of the net decrease in GHG emissions within the EU (EEA, 2020) (see Table 2.1).

Four primary sectors contribute to GHG emissions trends observed in Europe: energy (both use and supply) (60%), transport (19%), agriculture (10%), and waste management (3%) (EEA, 2011). The following subsections provide further details on these sources of climate change:

2.2.1 ENERGY CONSUMPTION

The consumption and production of energy through the use of fossil fuels has been a major driver of GHG emissions in Europe, particularly carbon dioxide (CO_2) emissions (EEA, 2007). Between 1990 and 2008, EU countries emitted approximately 28,000 million tons of GHGs (28,000 Mt $CO_{2\text{-equivalent}}$) into the atmosphere as a result of direct energy use (EEA, 2011). Direct energy use refers to the combustion of fuel products, while indirect use includes fuels used in the manufacture of vehicles, machinery, appliances, consumable goods, and buildings (EEA, 2011). It is important to note that European countries exhibit varying patterns of energy-related CO_2 emissions. According to EEA data on CO_2 emissions from energy use between 1990 and 2008, Germany, the United Kingdom, France, Italy, Spain, and Poland were the highest CO_2 emitters during this period.

2.2.2 AGRICULTURE ACTIVITIES

According to the United Nations Framework Convention on Climate Change (UNFCCC), GHG emissions from the agriculture sector encompass emissions resulting from agricultural activities such as livestock farming, rice cultivation, and synthetic

TABLE 2.1
Change in Total Greenhouse Gas (GHG) Emissions Over 1990–1980 in the European Countries. The Year 1990 Was Used as the Reference Year to Show Trends in GHG Emissions

	Total GHG Emissions in 2018 (MtCO$_{2e}$)	Change in Total GHG Emissions Over 1990–2018 (MtCO$_{2e}$)	Change in Total GHG Emissions over 1990–2018 (Percentage Change) (%)
Austria	81.5	2.1	2.7
Belgium	123.6	−25.9	−17.3
Bulgaria	58.6	−43.9	−42.8
Croatia	24.4	−8.0	−24.8
Cyprus	9.9	4.3	53.8
Czechia	129.4	−70.2	−35.2
Denmark	51.3	−21.3	−29.3
Estonia	20.2	−20.2	−50.0
Finland	58.8	−13.4	−18.6
France	462.8	−94.1	−16.9
Germany	888.7	−372.9	−29.6
Greece	96.1	−9.7	−9.2
Hungary	64.1	−30.4	−32.2
Ireland	64.2	7.7	13.6
Italy	439.3	−81.1	−15.6
Latvia	12.2	−14.4	−54.1
Lithuania	20.6	−27.8	−57.4
Luxembourg	12.4	−0.8	−5.8
Malta	2.7	−0.1	−3.9
The Netherlands	200.5	−25.8	−11.4
Poland	415.9	−59.9	−12.6
Portugal	71.6	11.4	18.9
Romania	116.5	−132.3	−53.2
Slovakia	43.5	−30.1	−40.8
Slovenia	17.6	−1.1	−5.7
Spain	352.2	58.1	19.7
Sweden	54.6	−17.9	−24.7
The United Kingdom	498.7	−311.0	−38.4
EU-27 and the United Kingdom	4,391.8	−1,329.5	−23.2
Total	3,893.1	−1,018.5	−20.7

Source: EEA (2020).

fertilizer use. Additionally, land use and land use changes such as deforestation and forest degradation, as well as forestry activities like wood harvesting are also included. In contrast to the energy sector where CO_2 emissions predominate, methane (CH_4), ammonia (NH_3), and nitrous oxide (N_2O) are the primary GHGs emitted by the agriculture sector (Mohammed et al., 2020). Rice cultivation is a significant global source of CH_4; however, it is infrequently practiced in Europe. As such, animal production (CH_4, N_2O, NH_3) and fertilizer use and production (N_2O and NH_3) are often the main sources of GHG emissions associated with agriculture in Europe (Brink et al., 2000; Mohammed et al., 2020).

In 1995, GHG emissions from agriculture in Europe amounted to approximately 470 Tg CO_{2e}, representing 11% of total GHG emissions and making agriculture the second-largest emitting sector after energy (Freibauer, 2003). The countries with the highest agricultural GHG emissions were France, Germany, the United Kingdom, Spain, and Poland (EEA, 2022). This high level of emissions can be attributed to agricultural overproduction of cereals, milk, and meat in the early 1980s (Freibauer and Kaltschmitt, 2001). However, between 1995 and 2005, there was a significant decrease (about 20%) in agricultural GHG emissions at the EU level due to successful implementation of various agricultural policies such as the Common Agricultural Policy (CAP) and Nitrates Directives across all EU countries (Van Grinsven et al., 2016; Heyl et al., 2020). From 2005 to 2020, there was no significant change in agricultural GHG emissions at the EU level; however, country-level trends varied significantly (EEA, 2022). For example, Croatia, Greece, Malta, and Romania experienced decreasing trends while Bulgaria, Estonia, Hungary, and Latvia saw increasing trends in agricultural GHG emissions during this period.

Agricultural GHG emissions in the EU are projected to decrease by 25% between 2015 and 2030 in order to help achieve climate neutrality by 2050 (Meessen et al., 2020). To reach this goal, urgent and decisive action must be taken including reducing emissions from livestock, increasing soil carbon sequestration and enhancing financing for the CAP (Verschuuren, 2022).

2.2.3 TRANSPORT

In the EU, in 2016, GHG emissions from the transport sector (including domestic transport, international aviation, and international maritime) reached approximately 0.9 Gt CO_{2e} with over 70% of these emissions originating from road transport (EEA, 2017; Plötz et al., 2021). Between 2017 and 2019, GHG emissions from the EU's transport sector grew steadily. However, in 2020, there was a significant decrease in emissions due to reduced mobility and transport activities during the COVID-19 pandemic (Kareinen et al., 2022). Without adequate measures, an increase in transport emissions could be observed in the coming years. In addition to the decrease in GHG emissions during the COVID-19 pandemic (Kareinen et al., 2022), further reductions are necessary to meet the EU's objective of reducing GHG emissions from the transport sector by 90% below 1990 levels by 2050 (Kareinen et al., 2022). The EU's Sustainable and Smart Mobility Strategy established in 2020 provides a clear path toward achieving this reduction through its 82 initiatives ten milestones and ten concrete actions aimed at decreasing GHG emissions from the EU's transport sector by 90% below 1990 levels by 2050 (Gkoumas et al., 2021).

2.2.4 WASTE MANAGEMENT

In recent decades, economic growth urbanization and industrialization have led to an increase in waste generation in the EU resulting in a rise in GHG emissions from waste management to critical levels (IPCC, 2013). In 1994, total GHG emissions from waste management in the EU reached a record 240 Mt CO_{2e} with over 50% of these emissions originating from solid waste disposal (EEA, 2011). When solid waste is disposed of in landfills, it decomposes and produces small amounts of CH_4. Under anaerobic conditions, methanogenic activity can cause large amounts of CH_4 to be emitted for several years even after landfills are closed (Chynoweth et al., 2001). CH_4 is a potent GHG with a global warming potential 28 times higher than that of CO_2 over a 100-year period (IPCC, 2013). However, between 1995 and 2017, emissions decreased dramatically by 42% due to the positive impact of EU legislation on waste management over the past 30 years (IPCC, 2013). This legislation includes mandatory policies and strategies aimed at reducing landfilling through recycling. As more waste is recycled, less is landfilled leading to reduced GHG emissions from waste. The long-term goal of this legislation is to transform Europe into a recycling society and reduce GHG emissions from waste management by 95 Mt by 2050 (Capros et al., 2013).

2.3 PAST AND FUTURE PROGRESS OF CLIMATE CHANGE IN EUROPE

Regardless of future GHG emissions scenarios (low, medium, or high), climate change will continue to impact EU countries in various ways due to the long atmospheric lifetime of GHGs and their continued accumulation (IPCC, 2013). Several research programs and projects such as the European Climate Prediction project ENSEMBLES Copernicus Earth Observation Programme and EURO-CORDEX have been conducted in recent years to produce accurate climate projections and predictions for European countries up to 2050. The key findings from these programs and projects can be summarized as follows:

1. All European countries will experience warming by 2050 with the highest warming expected in Mediterranean countries.
2. Precipitation patterns will change significantly with Northern Europe experiencing increased winter precipitation while Southern Europe will see decreased summer precipitation.
3. The frequency and intensity of flood events are projected to increase in all European countries except for those in the Mediterranean region.
4. Sea levels will rise in all European countries except for those bordering the Baltic Sea at rates close to or exceeding global mean sea level.

The following subsections provide further details on these forms of climate change:

2.3.1 CHANGE IN AIR TEMPERATURE

According to the Copernicus Climate Change Service (2022), the average air temperature in Europe increased at a rate of 0.05°C per year between 1991 and 2020, using the period from 1850 to 1900 as a reference. This rate is more than twice the global average of 0.02°C per year (Copernicus Climate Change Service, 2022). The year 2016 was recorded as the warmest year on record with a temperature rise of approximately 0.44°C above the average for the period from 1991 to 2020. It should be noted that there were significant regional differences in temperature increase rates during this period. Southern European countries such as France, Spain, and Italy experienced higher warming rates (>0.05°C per year) while Northern Europe experienced lower warming rates (<0.05°C per year). As a direct consequence of this warming trend, various forms of climate change such as ice melting, droughts, heatwaves, and floods have occurred in several European countries. For instance, since the 1980s, large masses of glaciers in the European Alps have been lost due to rapid warming observed in that region since the late 1970s (Grunewald and Scheithauer, 2010). Projections indicate that temperatures will continue to increase across all European countries between 2023 and 2050 at a rate similar to past observations regardless of future GHG emissions scenarios (Copernicus Climate Change Service, 2022). The highest warming is expected to occur in Mediterranean European countries (Copernicus Climate Change Service, 2022). Furthermore, there will be notable differences between summer and winter temperatures. Eastern Europe will experience the greatest changes during winter while northern Europe will have less cold winters. In contrast, Mediterranean countries are projected to have the highest temperature increases during summer (Carvalho et al., 2021). Based on temperature projections, it has been estimated that the volume of European glaciers will decrease significantly by 2010 (Copernicus Climate Change Service, 2022). Under a moderate GHG emission scenario (RCP4.5), the volume is projected to decrease by 22 to 84% compared to their current volume. Under a high emission scenario (RCP8.5), the decrease is projected to be between 38 and 89% (Marzeion et al., 2012; Radić et al., 2014).

2.3.2 CHANGE IN PRECIPITATION PATTERNS

According to the IPCC (2013), changes in precipitation patterns have manifested in two main forms in Europe over the past few decades.

- *Form 1: Change in annual precipitation amount*
  Compared to its amount in 1960, annual precipitation has increased at an average rate of 7 mm per year in Northeastern and Northwestern Europe and at an average rate of 9 mm per year in some areas of Southern Europe. In Southern Europe, the average amount of summer precipitation has decreased by approximately 2 mm per year, whereas in some areas of Northern Europe it has increased by approximately 1.8 mm per year (EEA, 2017).

• *Form 2: Change in frequency of heavy precipitation events*
  Using the period from 1951 to 1980 as a reference, the number of heavy precipitation events over the period from 1981 to 2013 at the European level has increased by 45% (Fischer and Knutti, 2016) as a direct result of rapid warming trends and increases in atmospheric water vapor. This increase has led to more severe flooding events with critical effects on human populations and ecosystems across all European areas, particularly along coastal regions. For example, a flood event that occurred over a period of 48 hours (July 14–15, 2021) in many parts of Germany, Belgium, and the Netherlands resulted in the loss of 180 human lives and significant damage to infrastructure (Lehmkuhl et al., 2022). According to Blöschl et al. (2019) and Lehmkuhl et al. (2022), in North-Western Europe, a typical 100-year flood event in 1960 has become a 50- to 80-year flood event. More frequent extreme flood events similar to the one that occurred in July 2021 are expected during the next few decades (Lehmkuhl et al., 2022). It is important to note that the number of heavy precipitation events varied significantly by region. Studies (Łupikasza, 2017; Tramblay and Somot, 2018) have identified distinct differences between Northern and Southern Europe: an increasing trend in the North and a decreasing trend in the South. This strong regional variation is caused by the complex interactions of both local factors (e.g., topography) and atmospheric circulation factors that trigger precipitation formation (Łupikasza, 2017).

Based on regional climate models from the EURO-CORDEX-0.11° project (http://www.euro-cordex.net/) and robust nonparametric statistics, several studies (e.g., Jacob et al., 2014; Prein et al., 2015; Zittis et al., 2021) have investigated future trends of precipitation in Europe. These studies have revealed that precipitation during the next few decades will vary substantially across regions and seasons. Generally, annual precipitation is projected to increase in Northern Europe and decrease in Southern Europe (EEA, 2017). The projected decrease in Southern Europe is strongest during the summer (EEA, 2017). Heavy precipitation events and their associated severe flood events are projected to increase at global warming levels exceeding 1.5°C in all European areas except for Mediterranean regions (EEA, 2017). Furthermore, some recent studies (e.g., Vousdoukas et al., 2018) have revealed that the costs of severe flood events may increase by up to three orders of magnitude by 2100 depending on GHG emissions scenarios.

2.3.3 Sea Level Rise

Sea level rise (SLR) has received significant attention over the past few years as an alarming indicator of climate change. Many European studies have investigated and reported on its past changes (e.g., Marcos et al., 2009; Weisse et al., 2014) and future trends (e.g., Vousdoukas et al., 2017; Calafat et al., 2022). These studies have shown that all coastal European areas have experienced an increase in SLR. In the majority of coastal European areas, sea level has risen at an average rate of about

2 to 4 mm per year over the period from 1993 to 2021, which is close to the global sea level rise estimated at 3.2 mm per year (WCRP Global Sea Level Budget Group, 2018; Ablain et al., 2019; Copernicus Climate Change Service, 2022). Only a few areas, such as some regions around the Mediterranean Sea, have shown a slight decrease in sea level (Copernicus Climate Change Service, 2022). As a direct result of SLR, most European coastal areas have experienced an increase in severe flood events (Copernicus Climate Change Service, 2022). The increase in SLR associated with coastal flood events is currently threatening coastal ecosystems, soil and water resources, infrastructure and human lives in all European coastal areas (Copernicus Climate Change Service, 2022). Sea level along most European coastal areas over the next few decades is projected to continue to rise at an increasing rate similar to today's rate with the exception of the northern Baltic Sea and northern Atlantic coast which are experiencing large land rises due to post-glacial rebound (IPCC, 2021). This means that more coastal floods should be expected (IPCC, 2021). In Europe, about 200 million people are currently exposed to SLR, and this number may increase significantly over the next few decades (Neumann et al., 2015). Therefore, it is urgently necessary to protect these people against SLR and its adverse effects through the implementation of adequate protection strategies. In this context, existing European policies such as the Integrated Coastal Zone Management (ICZM), Flood Risk Directive (FRD), Water Framework Directive (WFD), and European Green Deal (EGD) are particularly important and helpful in developing these strategies (e.g., Nixon, 2015; Bisaro et al., 2020).

2.3.4 HEATWAVES AND DROUGHTS

Since the beginning of the year 2000, heatwaves and droughts have become more frequent in Europe due to rising temperatures (Ionita et al., 2022). Several extreme heatwave events have occurred, including the Western European heatwave of 2003 which resulted in 70,000 deaths (Robine et al., 2008) and the Eastern European heatwave of 2010, which resulted in 55,000 deaths (Barriopedro et al., 2011). From 2017 to 2021, more than half of Europe was affected by extreme heatwave and drought events (Ionita et al., 2022), causing critical damage to people, economy, agriculture, transport, water supply, biodiversity, and infrastructure (Stahl et al., 2016). Droughts during the period from 1981 to 2010 caused approximately €9 billion in damages per year to the economies of EU countries and the United Kingdom (Naumann et al., 2021). Ionita and Nagavciuc (2021) analyzed drought trends across Europe from 1901 to 2019 using different drought indices and found that Central Europe and the Mediterranean region are becoming drier due to an increase in potential evapotranspiration and mean air temperature while North Europe is becoming wetter. Previous studies (Naumann et al., 2018; Spinoni et al., 2018; Cook et al., 2020) have projected an increase in the number of heatwave days in Europe under all GHG emission scenarios with greater increases expected for Southern Europe. Hotter heatwave events are also expected for many European cities. This suggests that European cities must prepare for this potential situation as both infrastructure and populace are not adapted to extreme heatwave events.

2.4   TARGET CLIMATE IN EUROPE BY 2050

By 2050, the European Union (EU) countries aim to limit global warming to below 1.5°C and achieve climate neutrality, defined as an economy with net-zero GHG emissions (European Commission, 2020). This goal is part of the EU's long-term strategy and European Green Deal (European Commission, 2019). To achieve this goal, EU countries (including the United Kingdom) have committed to stringent climate policies and measures over the past three decades that have significantly reduced GHG emissions. From 1990 to 2020, GHG emissions in EU countries decreased by approximately 31% (EEA, 2021). These measures include an increase in renewable energy use, a switch from coal to gas for power generation, improvements in energy efficiency and structural changes in EU economies (Figure 2.1) (EEA, 2021). For the period from 2023 to 2050, EU countries have decided to continue reducing GHG emissions and implement additional measures until net-zero emissions are achieved by 2050 (European Commission, 2020). This will require large-scale cooperation among involved actors, significant investments in sustainable environment and new regulatory frameworks and norms (Burke and Stephens, 2018). This means that all economic sectors, including energy use and production, agriculture, transport, forestry, land use, and waste management, will need to contribute to achieving net-zero GHG emissions at the European level by 2050 (European Commission, 2020).

While the objective for the next few years is clear (i.e., further reduction of GHG emissions), the pathway to climate neutrality is not (Kleanthis et al., 2022). Previous studies (e.g., Santos, 2017; EEA, 2021; Plötz et al. 2021; Kleanthis et al., 2022) have identified many challenges to achieving the target European climate by 2050. Some of these challenges include:

- Although EU countries share a common climate policy framework, circumstances such as energy use, geopolitical history, infrastructure, and protection options differ significantly among them. For instance, some countries possess abundant natural resources while others rely on importing fossil fuels (IEA, 2021; Kleanthis et al., 2022). As a result, countries will require

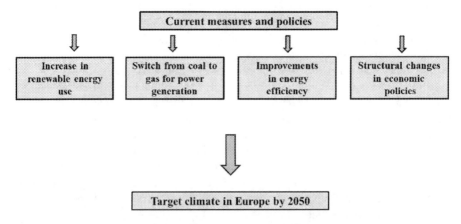

FIGURE 2.1   Measures and policies adapted by the European Union (EU) countries to achieve climate neutrality by 2050.

different pathways to achieve climate neutrality. A failure to account for these differences may pose a risk that Europe will not become the world's first carbon-neutral bloc by 2050 and meet the Paris Climate Agreement.

- European countries are currently considering various agricultural policies to promote farming practices (e.g., conservation tillage, organic material use, and tree planting) and land management options (e.g., producing renewable energy on agricultural land) that help reduce GHG emissions. While these practices and options are currently leading to significant increases in the amount of carbon removed from the atmosphere and stored in soil, several gaps must be addressed to sustainably maintain and enhance this trend. For example, proposed farming practices and land management options can affect farm profits. If agricultural policies aim to involve farmers voluntarily, they may need to include payment mechanisms to incentivize farmers to adopt relevant practices and options (Horowitz and Jessica, 2010).
- Unlike GHG emissions from other sectors, GHG emissions from the transport sector continue to increase. As a result, achieving GHG neutrality in transport appears particularly challenging (Plötz et al., 2021). This task is complex and difficult and requires significant effort due to the existence of many challenges and barriers. The main barriers include the high cost of clean technologies compared to carbon-intensive technologies, a lack of clear plans for future transport taxation and the fact that many European cities still lack sustainable transportation systems (e.g., Santos, 2017). According to Plötz et al. (2021), achieving net-zero carbon transport by 2050 is feasible but current transport policies must be strengthened.

2.5   CONCLUSIONS

This chapter examines Europe's climate change, particularly its past trends and key projections for 2050. The chapter demonstrates that climate change is affecting all European regions in various ways such as warming temperatures, altered precipitation patterns, increased heatwave and drought days, and more intense and frequent extreme floods along European coastlines in both the short and long term. Climate projections indicate that without significant action to reduce GHG emissions in the coming years, all European regions will face undesirable consequences of climate change by 2050. As this chapter highlights, a wide range of policies and strategies are available to European policymakers to help them take decisive steps toward further reducing GHG emissions over the next few years. In fact, many European countries have already reduced their GHG emissions due to a variety of factors including the positive impacts of EU policies; increased use of renewable energy; a switch from coal to gas for power generation; improvements in energy efficiency; and structural changes in EU economies. To achieve the goal of net-zero GHG emissions by 2050 and reach climate neutrality, it is imperative to harmonize European Union policies with those targeting GHG emission sources. The successful transition to carbon neutrality in Europe hinges on addressing key challenges such as enhancing soil carbon sequestration in agriculture and developing viable options for reducing GHG emissions from transportation in the coming years.

3 Africa
Past and Future Trends of Greenhouse Gas Emissions and Climate Change Indicators by 2050

3.1 INTRODUCTION

Africa is the world's second-largest continent, covering approximately one-fifth of Earth's total land surface. The continent is home to around 1.4 billion people, 50% of whom lack access to electricity (IEA, 2019), and 34% lived below the poverty line in 2019 (UNCTAD, 2021). Economic growth and food production in Africa have remained lower than the rest of the world (e.g., Bjornlund et al., 2020). As a result, Africa has contributed minimally to global greenhouse gas (GHG) emissions. In 2017, carbon dioxide (CO_2) emissions from Africa were 1,185 $MtCO_2$, representing only 4% of global emissions (Ayompe et al., 2020). However, since the 1970s, Africa has experienced the effects of global GHG emissions in various forms of climate change such as warming temperatures, droughts, changes in precipitation patterns, and rising sea levels (Ofori et al., 2021). Due to limited economies, high population growth and low adaptive capacity to climate change, these effects have had severe impacts on people, economy, agriculture infrastructure, and ecosystems over the past few decades (Niang et al., 2014). For instance, in East Africa, about 226 drought events occurred between 1979 and 2014 resulting in significant socio-economic impacts including desertification, soil salinization, crop failures, livestock deaths, and human casualties (IPCC, 2007; Zhao and Dai, 2015; Liu et al., 2022). Thus, Africa is considered one of the world's most vulnerable continents (Boko et al., 2007).

In recent years, climate change in Africa has received increased attention. Studies utilizing multiple data sources and climate models (e.g., Vizy and Cook, 2012; Haensler et al., 2013; Ongoma et al., 2018; Almazroui et al., 2020a) have examined past climate trends and projected future changes and their potential impacts. These studies have identified global warming as the primary driver of climate change, with additional contributing factors including population growth, oil production, and low levels of renewable energy production and use (Amoah et al., 2020; Almazroui et al., 2020a). These factors are expected to increase GHG emissions and exacerbate the effects of climate change across the continent (IPCC, 2022). The human impacts are projected to be severe, with increased mortality from malnutrition, diarrhea,

DOI: 10.1201/9781003404194-4

malaria, and dengue fever if effective mitigation measures are not implemented (Byass, 2009). The United Nations Environment Programme (UNEP) estimates that the cost of such measures in Africa is approximately $50 billion per year (IPCC, 2022). Given these projections, it is crucial for policymakers to improve their understanding of past and future climate trends and GHG emissions in order to make informed adaptation decisions. Achieving climate neutrality by 2050 (i.e., net-zero GHG emissions) is essential for mitigating the potential impacts of climate change on Africa (IPCC, 2022). According to Al-Zu'bi et al. (2022), Africa has the potential to play a leading role in promoting climate change adaptation, disaster risk reduction, and sustainable development. In this context, this chapter aims to (1) analyze past and future trends of GHG emissions in Africa; (2) assess historical and projected climate change trends in Africa; and (3) evaluate Africa's contribution to achieving global net-zero GHG emissions by 2050.

3.2   PAST AND FUTURE TRENDS OF GHG EMISSIONS IN AFRICA

According to the IPCC (2022), Africa's 54 countries have contributed just 7% of global GHG emissions between 1850 and 2019, with this percentage dropping to only 4% in 2017 (Ayompe et al., 2020). Furthermore, Ongoma et al. (2023) reported a decrease in GHG emissions on the continent during 2020–2021 due to the COVID-19 pandemic and associated policies such as travel restrictions and reduced energy use at workplaces during stay-at-home orders. Greenhouse gas emissions vary significantly across African countries. South Africa, Egypt, Algeria, Nigeria, Morocco, and Libya were the leading GHG emitters in Africa in 2019 with respective emissions of 439, 249, 171, 115, 71, and 56 Megatons (Mt) of CO_2. These six countries accounted for over half of the continent's total emissions. This high percentage can be attributed to their economies being heavily reliant on fossil fuel combustion which releases large amounts of CO_2 (e.g., Blignaut et al., 2005).

A study by Ayompe et al. (2020) found that factors such as population growth, continued oil production, and low levels of renewable energy production may increase Africa's CO_2 emissions by 30% from 1,185 $MtCO_2$ in 2017 to 1,545 $MtCO_2$ in 2030. The study also projected that South Africa, Egypt, Algeria, and Nigeria would remain the leading GHG emitters in Africa in 2030. Therefore, it is imperative that effective mitigation measures should be implemented as soon as possible to help the continent contribute to achieving global net-zero GHG emissions by 2050 (IPCC, 2022). In addition to fossil fuel combustion, which accounted for 35% of GHG emissions in Africa in 2016, approximately 57% of total emissions came from the Agriculture, Forestry and Other Land Use (AFOLU) sectors which represent a significant portion of many African economies (Osman-Elasha and de Velasco, 2021). Agricultural emissions primarily consist of methane (CH_4) from enteric fermentation (e.g., cattle and sheep digestion) and nitrous oxide (N_2O) from manure and synthetic fertilizer applications (Steinfeld and Wassenaar, 2007). Greenhouse gas emissions from AFOLU sectors in Africa are among the fastest growing globally (IPCC, 2022). According to the FAO (2015), Africa is the third largest emitter of agricultural GHGs after Asia and the Americas. A review by Tongwane and Moeletsi (2018) found that GHG emissions from the agricultural sector in Africa increased at an average annual rate of about 3.1% between 1994 and 2014 – more than double the global average rate. However, it should be noted that agricultural GHG emissions vary significantly across African countries. According to data from FAO (2015), Southern and North African countries have lower levels of agricultural GHG emissions while Eastern African countries have higher levels. In fact, about a third of Africa's total agricultural emissions come from Eastern African countries (Tongwane and Moeletsi, 2018).

Agricultural GHG emissions in Africa also vary significantly by category. Generally, a large proportion (>50%) of agricultural emissions come from enteric fermentation while smaller amounts come from the management of crop residues and organic soils (Tongwane and Moeletsi, 2018). It is important to note that promoting agricultural improvement measures such as agroforestry, soil and water conservation, and efficient use of nitrogen can help reduce future agricultural GHG emissions in Africa and mitigate climate change (Kim et al., 2021a).

3.3   CLIMATE CHANGE FORMS IN AFRICA

Research on the effects of climate change on African countries has rapidly evolved in recent years due to their vulnerability to climate fluctuations (Bouramdane, 2023). The latest Coupled Model Intercomparison Project (CMIP6) models and downscaled models provide accurate climate information for the entire African continent with acceptable confidence (Almazroui et al., 2020a). Studies using these models have revealed findings suitable for decision-making (e.g., Almazroui et al., 2020a,b; Quenum et al., 2021; Choi et al., 2022; Bouramdane, 2023). Major findings include significant variation in climate change across Africa due to its large latitudinal extent. Some areas are becoming wetter while others are getting drier and hotter (Almazroui et al., 2020a). Depending on future GHG emission scenarios, Africa's average annual temperature is projected to increase by 1.2°C to 1.8°C by 2050 (Almazroui et al., 2020a). Northern and Southern Africa will face decreased precipitation while central parts will experience increased precipitation over the next decades (2030–2050) (Almazroui et al., 2020a). Sea levels are projected to rise along African coastlines and heatwaves and droughts will become more frequent (e.g., Ajibola et al., 2022; Ayugi et al., 2022) (Figure 3.1). The following subsections provide details about these forms of climate change.

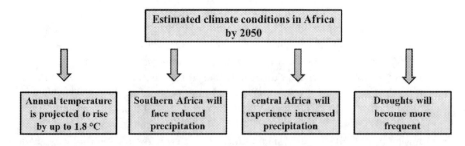

FIGURE 3.1   Estimated climate conditions in Africa by 2050.

3.3.1 INCREASING TEMPERATURES

Over the past five decades, Africa's average air temperature has risen by approximately 0.2°C per decade (e.g., Collins, 2011; Stern et al., 2011; Misganaw Engdaw et al., 2021). However, this warming varies significantly across the continent. Southern Africa experienced the lowest temperature rise (about 0.15°C per decade) while Northern Africa observed the highest rise (0.27 to 0.41°C per decade) (Misganaw Engdaw et al., 2021), almost twice the global rate. In recent years, many African regions have experienced extreme high daily temperatures due to this warming trend (Nangombe et al., 2019). For example, Egypt's Luxor region faced extreme high temperatures of about 47.6°C and South Africa's Vredendal region experienced extreme high temperatures of 48.4°C (WMO, 2016). Djibouti's hot days (temperature >45°C) have become 15 times more frequent over the period from 1966 to 2011 than in the past (Ozer and Mahamoud, 2013). Recent CMIP6 projections indicate that if adequate climate change mitigation measures are not undertaken, Africa will warm faster than global mean temperature (e.g., Almazroui et al., 2020a,b). Using a reference period of 1981–2010, Almazroui et al. (2020a) revealed that by 2050 Africa's average annual temperature is projected to rise by 1.2°C under low GHG emissions scenarios (SSP1-2.6), by 1.5°C under moderate scenarios (SSP2-4.5), and by 1.8°C under high scenarios (SSP5-8.5). However, this future warming varies largely across the continent with North Africa expected to face the highest temperature rise and East Africa expected to face the lowest (Nangombe et al., 2019). For West Africa, it is expected that temperatures will mainly increase in summer season by 2050 (Diallo et al., 2012). The future warming trend across Africa is expected to result in drastic increases in the frequency of very hot days (Niang et al., 2014). This warming will put additional pressure on ecosystems, agricultural systems, and water resources, particularly in arid and semiarid African regions. As such, there is an urgent need to promote effective ways to address the issue of African warming.

3.3.2 CHANGE IN PRECIPITATION PATTERNS AND FLOODS

Previous studies indicate that many African countries have experienced notable changes in precipitation patterns over recent decades (e.g., Collier et al., 2008; Shongwe et al., 2011; Ongoma and Chen, 2017). From 1961 to 2015, Africa's precipitation decreased by a rate of 8 mm per decade (Onyutha, 2021). However, due to the poor density of in-situ precipitation observations in some African regions such as the Sahara and central equatorial Africa (Nicholson et al., 2018), it is difficult to draw detailed conclusions about precipitation trends for the entire continent. The Global Climate Observing System (GCOS) and the World Meteorological Organization (WMO) identified that Africa has the weakest density of meteorological stations globally (e.g., López-Ballesteros et al., 2018; Merbold et al., 2021). Despite this, some studies have provided a good description of past trends in seasonal, annual, and decadal rainfall over Northern, Western, Eastern, Central, and Southern Africa (e.g., Lyon and Dewitt, 2012; Nicholson et al., 2018; Gaetani et al., 2020). These trends are summarized in Table 3.1.

Due to global warming, Africa has experienced frequent intense precipitation events over the past decades that are often associated with deadly floods (Tramblay et al., 2020).

TABLE 3.1
Change in Precipitation Patterns Over Northern, Western, Eastern, Central, and Southern Africa during the Past Few Decades

Region	Change in Precipitation Patterns	Key Reference(s)
Northern Africa	Annual precipitation decreased remarkably since the 1980s.	Nicholson et al. (2018)
Western Africa	The region showed a recovery in rainfall amounts since the 1980s, after several years of shortage.	Gaetani et al. (2020)
Eastern Africa	Over the past few decades, the annual rains have been reducing continuously.	Nicholson (2017)
Central Africa	Rainfall remained relatively abundant, although it seems to has decreased since the 1970s. Rainfall has remained relatively abundant, although it appears to have decreased since the 1970s.	de Wasseige et al. (2015)
Southern Africa	Over the past few decades, most of the Southern African areas experienced an important decrease in annual and seasonal rainfall magnitude and rainfall duration. Over the past few decades, many areas in Southern Africa have experienced a significant decrease in both the magnitude and duration of annual and seasonal rainfall.	Jury (2013) Daron (2014)

From 1950 to 2019, thousands of intense daily precipitation events occurred in Africa causing more than 27,000 deaths and affecting about 82 million people while resulting in significant economic damage (Tramblay et al., 2020). Heavy flood events frequently occurred in Western Africa and North/South Africa (Tramblay et al., 2022). Recent CMIP6 projections indicate that Northern and Southern Africa are expected to experience a reduction in precipitation while Central Africa is expected to see an increase over the coming decades under low (SSP1-2.6), moderate (SSP2-4.5), and high (SSP5-8.5) GHG emissions scenarios (e.g., Almazroui et al., 2020a). Kendon et al. (2019) also showed that tropical Africa is expected to face more severe precipitation events while experiencing a decrease in wet-season rainfall. This means that the risk of extreme floods is very likely for this region. As such, adequate safety measures should be promoted to avoid deadly floods which could be a key challenge for the entire African continent.

3.3.3 SEA LEVEL RISE AND COASTAL EROSION

Africa has an extensive coastline of approximately 39,000 km (McLachlan, 2019), encompassing roughly 320 coastal cities such as Lagos (Nigeria), Dar es Salaam (Tanzania), Alexandria (Egypt), and Cape Town (South Africa) (Hinkel et al., 2012). These coastal cities were home to over 54 million people in 2000 and continue to attract populations due to their significance in industrial, commercial, and agricultural sectors (Neumann et al., 2015). Consequently, sea level rise and its associated risks of flooding and inundation may have severe impacts on African populations, transportation systems, agriculture, ecosystems, and water resources within these regions (IPCC, 2021). Several coastal African cities are particularly vulnerable to sea level rise resulting in increased rates of coastal erosion along sandy coasts (de la Vega-Leinert et al., 2000). Benin, Egypt, Mauritania, and Tunisia rank among the top ten African countries most vulnerable to sea level rise (Dasgupta et al., 2009; Hinkel et al., 2012). However, long-term observations (>50 years) of sea level rise are limited in Africa compared to other continents making it difficult to draw detailed conclusions about trends over recent decades for the entire continent (Hinkel et al., 2012). According to the sixth assessment report of the IPCC (IPCC, 2021), over the past three decades, sea level rise in Africa has increased at a higher rate than the global average estimated at 3.4 mm per year. However, this rate varies significantly by country with some areas experiencing rates as low as 4.51 mm per year while others experience rates exceeding 20 mm per year (Sarr et al., 2021). According to the sixth assessment report of the IPCC, sea level rise is projected to continue around Africa over the coming decades, even if GHG concentrations stabilize. This will result in increased frequency and severity of coastal flooding and erosion in low-lying areas. For instance, Kebede et al. (2018) projected that sea level rise at the Gulf of Guinea will increase from 0.21 m in 2000 to 0.36 m by 2050. The lack of suitable coastal protection measures, such as drainage measures, could lead to catastrophic effects. Despite several coastal African cities experiencing significant coastal erosion due to sea level rise, few studies have quantified the relationship between coastal erosion and sea level rise rate. For example, Kebede et al. (2018) and Nyadzi et al. (2021) estimated that along the coast of Côte d'Ivoire, Ghana, Togo and Benin an overall coastal soil erosion rate of about 25 to 80 cm per year is expected by 2100 due to a sea level rise of 30 cm and 100 cm respectively.

3.3.4 Drought Events

Drought can be defined as a natural process resulting from a significant decrease in water availability due to reduced precipitation over an extended period of time, typically several months (Bayissa et al., 2015). Other factors such as increased temperature, evaporation and changes in cloudiness regimes can exacerbate drought conditions (Shilenje et al., 2019). Drought can have severe consequences for local populations including death of livestock and humans, shortage of drinking water, damage to agriculture, severe hunger, disruption of livelihoods, and significant economic loss. Several tools have been developed to detect and analyze drought occurrence in specific areas (IPCC, 2021). These tools will be discussed in detail at the end of this chapter along with a practical exercise on their use. In Africa, where millions of people experience water scarcity, drought is prevalent and considered one of the most dangerous natural disasters that can exacerbate existing water crises (El Kenawy et al., 2016). The Centre for Research on the Epidemiology of Disasters (CRED) recorded more than 290 severe drought events in Africa between 1900 and 2013 resulting in 847,100 deaths and affecting 362 million people (particularly children) causing economic damage estimated at approximately $2.9 billion USD (IPCC, 2021). For example, in 2011, the Great Horn of Africa experienced one of the world's most severe drought events resulting in widespread food shortages and loss of life (Shilenje et al., 2014). This catastrophic event was directly attributed to the failure of both the October-December "short rains" and March-May "long rains"

TABLE 3.2

Drought Events in Some African Countries and Their Visible Impacts Over the Last Two Decades

African Country	Drought Events	Visible Impacts	Key Reference
Ethiopia	The drought of 2015	The event affected more than 10 million people.	Mekonen and Berlie (2020)
Uganda	The droughts of 2005–2008	The droughts resulted in a drastic drop in Lake Victoria water levels by 1.2 m below the 1951–2006 average. The droughts caused a drastic drop in the water levels of Lake Victoria, falling 1.2 meters below the average level recorded between 1951 and 2006.	Awange et al. (2008)
Ghana	The drought of 2013–2014	The drought decreased significantly water amounts of many dams, which in turn reduced the hydropower generation capacity of the dams. The drought significantly decreased the water levels in many dams, reducing their capacity to generate hydropower.	UNCCD (2015)
Zimbabwe	The drought of 2015–2016	The drought affected food security of about 2 million people and caused 40 000 cattle deaths.	Soul et al. (2022)
South Africa	The drought of Eastern Cape province in 2019	The province was declared a drought disaster area due to its contribution to serious water shortages in many urban and rural areas. The province was declared a drought disaster area due to the severe water shortages it caused in many urban and rural areas.	Mahlalela et al. (2020)

during 2010 (Funk, 2011; Shilenje et al., 2019; Ongoma et al., 2015). Previous studies (e.g., Mekonen and Berlie, 2020; Abara et al., 2020) have shown that the frequency, duration and intensity of drought events in Africa have increased over the past two decades. These studies also indicate that droughts on the continent are worsening and that urgent policy and protective measures are needed. Protective measures against droughts in Africa can cost millions of dollars annually (Martey et al., 2020). Table 3.2 summarizes some recent examples of drought events in Africa and their impacts. Using data from the sixth phase of the Coupled Model Intercomparison Project (CMIP6), several recent studies (e.g., Ajibola et al., 2022; Ayugi et al., 2022) have investigated future changes in drought events over Africa. Under different future GHG emission scenarios (from low to high), these studies predict a significant increase in the frequency of drought recurrence over the coming decades due to climate change and rapid population growth in Africa. Therefore, there is an urgent need for early preparation for this potential situation. In this context, it is essential to promote research into future trends of droughts in Africa. The results of these future studies will be useful for identifying drought hotspots on the continent and developing appropriate adaptation and protective measures (Ayugi et al., 2022).

3.4   POTENTIAL CONTRIBUTION OF AFRICA
TO CLIMATE NEUTRALITY BY 2050

To limit global warming to below 1.5°C, GHG emissions must be reduced by 50% before 2030 and reach net-zero emission by 2050 (Levin, 2018). To achieve this goal of climate neutrality by 2050, the international community must promote climate change policies and sustainable development measures that can significantly reduce GHG emissions from developed countries and prevent large emissions from developing countries, particularly Africa, which contributes minimally to global GHG emissions (Ayompe et al., 2020). Africa has the potential to contribute to climate neutrality by 2050 through various GHG emission reduction pathways. Due to its unique emission profile, the GHG emission reduction pathways for African countries differ from those of developed countries (Bouchene et al., 2021). Previous studies (e.g., Jindal et al., 2008; Dyson, 2017; Momodu et al., 2022) have identified two key strategies for decarbonization in Africa as shown in Figure 3.2.

- Strategy 1: Transitioning to less carbon-intensive fuels and replacing polluting electricity sources with renewable energy sources, particularly in the power sector (Jordaan et al., 2017). The electricity sector must achieve full decarbonization by 2050 to meet the Paris Agreement's goal of limiting global temperature rise to 1.5°C (IPCC, 2018).
- Strategy 2: Increasing carbon sequestration. For instance, Africa has extensive degraded areas that can be utilized for developing carbon sinks through afforestation and reforestation initiatives (Jindal et al., 2008) and agroforestry (Kim et al., 2021b).

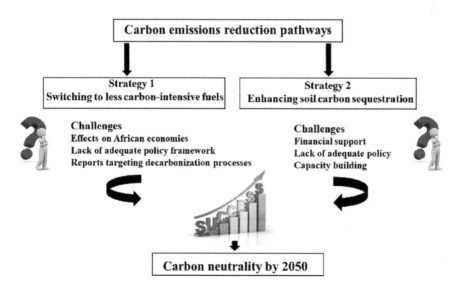

FIGURE 3.2   Greenhouse gas emissions reduction pathways in Africa and their associated challenges.

It is important to note that there are numerous challenges that may impede the success of these two strategies and diminish Africa's potential to achieve climate neutrality by 2050. First, climate neutrality policies may have significant impacts on African economies. As the rest of the world progresses toward net-zero emissions, Africa may experience a decline in global demand for fossil fuels and its exports could become less competitive (Momodu et al., 2022). Additionally, Röttgers and Anderson (2018) stated that inadequate policy frameworks are a major obstacle to promoting renewable energy sources in many African countries. Furthermore, Africa's limited utilization of renewable energy sources undermines its ability to build climate resilience; indeed, many African countries continue to invest in fossil fuel-based energy (Momodu et al., 2022). Moreover, there is a critical need for reports targeting decarbonization projects to accurately assess the continent's contribution to global emission reduction (Momodu et al., 2022). Enhanced policy environments that foster innovations in land planning, capacity building and financial support are also necessary in many African countries to promote carbon sequestration (Demenois et al., 2020). Addressing these challenges can transform "the transition to net-zero by 2050" into an opportunity for Africa to develop a low-carbon economy that protects and improves lives and livelihoods (Bouchene et al., 2021). In this context, identifying key tools that can accelerate the climate transition in Africa is essential.

3.5 CONCLUSIONS

This chapter of the book reviews the issue of climate change in Africa based on findings from numerous published studies. First, it reveals that although the continent is not a major contributor to global GHG emissions (<10% of global emissions), it has experienced various forms of climate change since the 1970s, including rising temperatures, droughts, changes in precipitation patterns and sea level rise. These changes may have significant impacts on African populations. Furthermore, climate projections indicate that the situation will worsen in the coming decades, necessitating international assistance to help Africa prepare for this critical situation through appropriate adaptation strategies. This is essential to protect African populations and their valuable ecosystems. The chapter also highlights that research on climate change has not been conducted in all African countries and further investigations targeting all countries are needed. Finally, Africa continues to attract researchers due to its potential to contribute to the global transition toward net-zero GHG emissions by 2050.

EXERCISE (WITH ANSWERS)

Question: How can droughts be detected in a given area?

Answer: To detect drought trends in a given area, it is essential to calculate the Standardized Precipitation-Evapotranspiration Index (SPEI). This statistical indicator identifies shortages or excesses of precipitation over time. It is one of the most widely used drought indices worldwide and is determined through a series of successive steps.

Step 1: Collect monthly temperature and rainfall (or precipitation) data for your study area over an extended period, typically 30 years or more.

Step 2: Calculate the monthly Potential Evapotranspiration (PET) using standard equations such as the Penman-Monteith equation or the Thornthwaite equation.

Step 3: Calculate the monthly water balance for each month by subtracting PET from rainfall.

Step 4: Convert the water balance into a standardized form by fitting a probability distribution (e.g., Gaussian distribution) to the data and transforming the values into standard normal variates.

Step 5: Compute SPEI values by accumulating standardized monthly water balance values over a specified time scale (e.g., 12 months) and multiplying accumulated values by the standard deviation of the fitted distribution.

Step 6: Interpret results. Generally, negative SPEI values indicate drought conditions in the study area while positive SPEI values indicate wet conditions.

It is worth noting that there are many software packages and online tools (e.g., R packages) that can facilitate steps 5 and 6.

4 Climate Change in Asia
Past Observations, Future Trends and Opportunities for Achieving Global Climate Neutrality by 2050

4.1 INTRODUCTION

Climate change is widely recognized as the most significant global environmental issue (IPCC, 2021), and many countries are currently experiencing significant changes in climate variables such as temperature and rainfall. In Asia, climate has also changed dramatically, with increases in temperature, frequent droughts and heatwaves, more deadly floods, and changes in precipitation patterns (Anbumozhi et al., 2012). Over the past few decades, these various aspects of climate change have had significant impacts on populations, economies, and ecosystems. For example, China has experienced an annual loss of approximately 2.3% of its Gross Domestic Product (GDP) due to climate disasters (Wang and Zheng, 2012). In India, even with efficient crop management measures in place, wheat yields decreased by about 5% between 1981 and 2009 due to rising temperatures (Gupta et al., 2017). Furthermore, severe flood events in Pakistan over the past six decades have resulted in more than 10,000 deaths and an estimated economic loss of US $30 billion (Shah et al., 2020).

According to previous studies (e.g., Rahman, 2013; Ab-Rahim and Xin-Di, 2016; Filimonova et al., 2022), increasing greenhouse gas (GHG) emissions in Asia can be attributed to several factors. The primary factor is the high consumption of fossil fuels, resulting in an estimated 17.7 billion metric tons of carbon dioxide (CO_2) emissions in 2021 and making Asia the world's largest CO_2 emitter (BP Statistical Review of World Energy, 2022). Other contributing factors include trade openness, rapid economic growth, dependence on agriculture, high population density, urbanization, and weak institutions in some countries. However, the intensity of these factors varies across Asia. For instance, China emitted approximately 10.5 billion metric tons of CO_2 in 2021 (Statistical Review of World Energy), making it the top emitter both in Asia and globally. In contrast, Sri Lanka emitted only about 0.02 billion metric tons.

Recent climate projections for Asian countries from the Coupled Model Intercomparison Project-6 (CMIP6) have garnered significant attention (e.g., Mishra et al., 2020; Shahid and Rahman 2021; Liu et al., 2021; Ali et al., 2022). These projections indicate that climate change in Asia is likely to intensify in the coming decades.

DOI: 10.1201/9781003404194-5

By 2050, large parts of the region may experience rising average temperatures, lethal heatwaves, extreme precipitation events, severe droughts, and critical sea level rise. Such trends could result in serious human health issues (including mortality), catastrophic damage to ecosystems and infrastructure and significant economic loss (IPCC, 2018). It is therefore crucial to better understand the changing climate over the coming decades to provide decision makers with valuable information, identify knowledge gaps, and develop effective adaptation strategies. Given its significant contribution to climate change, reducing GHG emissions in Asia is an urgent task that must be addressed in the medium term. According to Nishioka (2016), Asia is well-positioned to effectively address climate change and can achieve the global target of net-zero GHG emissions by 2050. To implement this transition, it is essential for Asian countries – particularly China – to align with international efforts and promote carbon policies through various actions and measures. In this context, this chapter aims to review published studies on Asian climate change. The specific objectives are to (1) provide a quantitative analysis of historical GHG emissions in Asia; (2) discuss long-term changes in climate over past and future decades; and (3) analyze Asia's key role in achieving global climate neutrality by 2050.

4.2   HISTORICAL GREENHOUSE GAS EMISSIONS IN ASIA

It is important to note that from the beginning of the 19th century until 2005, the United States was the world's largest emitter of GHGs. However, since 2006, Asian countries, led by China, have emerged as the largest GHG emitters due to industrialization and population growth (Friedrich and Damassa, 2014). In these countries, GHG emissions from various human activities (e.g., fossil fuel consumption and land use) have continuously increased since 1990, reaching their highest levels in 2019. Overall, Asia accounts for over 50% of global emissions (Azhgaliyeva and Rahut, 2022). According to Lamb et al. (2021), GHG emissions in Asia have three main sources: energy consumption, transport and agriculture, forestry and other land uses (AFOLU). The following subsections provide details about these sources:

4.2.1   GREENHOUSE GAS EMISSIONS FROM ENERGY CONSUMPTION

Energy use constitutes the largest share of total GHG emissions. Previous studies (e.g., Lu, 2017; Salahuddin et al., 2019; Wang et al., 2019) have shown that fossil fuel consumption – particularly coal, oil and natural gas – has intensified CO_2 emissions in Asian countries (especially China and India) over the past few decades. Lu (2017) found that a 1% increase in energy consumption in Asian countries can result in approximately a 0.8% increase in CO_2 emissions. High-polluting and energy-intensive industries such as cement production, power generation and petrochemical processes are the dominant sources of energy consumption and associated CO_2 emissions in Asia (e.g., National Bureau of Statistics, 2018). For instance, these industries resulted in approximately 8 billion tons of CO_2 emissions in China in 2012 – accounting for 91% of the country's total CO_2 emissions that year (Institute of Climate Change and Sustainable Development of Tsinghua University, 2022).

In 2020, lockdown measures implemented due to the COVID-19 pandemic played a positive role in decreasing CO_2 emissions in some Asian countries (Ray et al., 2022). However, not all Asian countries experienced a decrease: while India emitted 325 Mt less CO_2 than it did in 2019 (Ray et al., 2022), China saw an increase after ending its lockdown measures early (in April) (Liu et al., 2020). In 2021, Asia emitted approximately 17.7 billion metric tons of CO_2 – making it the world's top emitter (BP Statistical Review of World Energy report). It is important to note that CO_2 emissions vary significantly across the Asian region – particularly in relation to China and India. China alone contributed 59% of the region's emissions (approximately 10.5 billion metric tons), while India accounted for 14.3% (approximately 2.5 billion metric tons). Given the accelerated economic growth expected in the coming decades, emissions from Asia in 2050 are projected to double their 2005 levels if actions and measures are not taken toward achieving low-carbon societies. This means that reducing GHG emissions in Asia is crucial for transitioning to climate neutrality by 2050 (Masui et al., 2016).

4.2.2 GREENHOUSE GAS EMISSIONS FROM TRANSPORT

In many Asian countries, the transport sector is a primary contributor to CO_2 emissions. Rapid urbanization and motorization have led to significant increases in CO_2 emissions and concerns about climate change (IEA, 2019). For instance, CO_2 emissions from China's transport sector rose from approximately 123 Mt in 1995 to 670 Mt in 2012. Motor vehicles are the primary source of CO_2 emissions within the road transport subsector due to their reliance on fossil fuels (IEA, 2019). Studies have shown that motor vehicles can emit substantial amounts of CO_2; for example, Unzilatirrizqi et al. (2019) found that motor vehicles in Tegal City, Indonesia emitted approximately 1,520 Mg/hour.km of CO_2. Similarly, Kumar et al. (2022) reported that up to 90% of India's transportation-related CO_2 emissions originated from the road subsector. They also found that lockdown measures during the COVID-19 pandemic reduced road transport travel demand by 11% in 2020 compared to 2019 and shifted the trajectory of CO_2 emissions within this subsector. In addition to road transport, air transport also contributes significantly to CO_2 emissions (IEA, 2019). For example, Tarr et al. (2022) reported that New Zealand's air transport subsector emitted approximately 8.4 Mt of CO_2 in 2017. Given projected population growth in Asia over the coming decades and without effective measures toward low-carbon transport systems, it is expected that emissions from transport energy consumption will increase significantly by 2050 (ITF, 2019). Implementing measures such as improving fuel efficiency, promoting vehicle electrification, increasing the use of alternative fuels, and reducing travel demand could help mitigate the growth trend of CO_2 emissions from the transport sector (Kumar et al., 2022). These measures could accelerate Asia's progress toward achieving its climate objectives. Implementing measures such as improving fuel efficiency, promoting vehicle electrification, increasing the use of alternative fuels, and reducing travel demand could help mitigate the growth trend of CO_2 emissions from the transport sector (Kumar et al., 2022). These measures could accelerate Asia's progress toward achieving its climate objectives.

4.2.3 GREENHOUSE GAS EMISSIONS FROM AGRICULTURE, FORESTRY AND OTHER LAND USES

Globally, Agriculture, Forestry and Other Land Use (AFOLU) contributed to approximately 24% of total global GHG emissions in 2010 (IPCC, 2014). However, this percentage was significantly higher in Asian countries at 44%, due to the extensive cultivation of rice which is a major source of methane (CH_4) emissions (IPCC, 2014). In South Asian countries, this percentage can reach up to 67% (PRIMAP, 2016). Previous studies have shown that CH_4 and nitrous oxide (N_2O) are the primary GHGs emitted from the AFOLU sector (e.g., Chhabra et al. 2013; Zhang et al., 2021). Numerous studies have investigated GHG emissions from the AFOLU sector in Asia, and their main findings can be summarized as follows:

- First, GHG emissions from the AFOLU sector vary by region (Aryal, 2022).
- Second, since 90% of global rice production occurs in Asia, the region is considered the largest source of global CH_4 emissions (Kritee et al., 2018), with an estimated 87% of CH_4 emissions from rice cultivation originating from Asia (Aryal, 2022).
- Third, due to the intensive use of inorganic fertilizers, Asian countries are responsible for 70% of total fertilizer-related N_2O emissions (Lassaletta et al., 2014), with a significant portion resulting from the manufacturing and transportation of these fertilizers (e.g., Liang et al. 2021; Aryal, 2022).
- Fourth, burning crop residues on farms has been observed to contribute significantly to CO_2 emissions in many Asian countries, particularly India (e.g., Shen et al., 2021). For example, burning approximately 116 million tons of crop residue in India during the 2017/2018 season resulted in approximately 176 Tg of CO_2 emissions (Venkatramanan et al., 2021).
- Fifth, certain agricultural practices such as manure management and irrigation can emit large amounts of CH_4 and N_2O (Aryal, 2022).
- Sixth, high levels of deforestation and other land use changes have contributed significantly to GHG emissions in some Asian regions over the past few decades (Aryal, 2022).
- Finally, without effective control measures, global warming is expected to increase GHG emissions from the AFOLU sector in Asia (e.g., Yue et al. 2017; Aryal, 2022). However, there is significant potential to reduce GHG emissions from this sector if effective measures are implemented across all Asian countries. Agriculture is a critical sector in addressing climate change (Aryal, 2022).

4.3 LONG-TERM CHANGES IN CLIMATE OVER PAST DECADES

Several studies have investigated long-term changes in climate over past decades in Asia (e.g., Ren et al., 2021; Muhammad et al., 2021; Ngai et al., 2022). However, it has been revealed that most of these studies focused on climate change in the eastern part of Asia (especially China), while Central and North Asia received less attention. These studies have shown that the Asian region has experienced a significant warming trend, potentially increasing the risk of extreme climate events. In addition to this warming trend, the region has also undergone significant changes in precipitation patterns and an increase in the frequency of extreme climate events. Further details about these aspects of climate change are presented in the following subsections (Table 4.1):

TABLE 4.1
Observed Changes in Climate Variables Over Past Decades across the Asian Domain

Region	Change in Temperature	Change in Precipitation	Change in Sea Level	Extreme Climate Events
North Asia	Over 1950–2005, the region witnessed an overall increase in winter and spring temperatures by 5 to 8°C/century.	The region experienced a significant upward trend in precipitation for all seasons except summer over 1901–2016.	(-)	Since the mid-1970s, the region experienced more intense floods (IPCC, 2021).
South Asia	Studies revealed rising temperature trends over most countries of South Asia, which are more pronounced during winter than in summer (Naveendrakumar et al., 2019).	The region faced a significant downward trend in winter over 1901–2016.	Sea level around South Asia has raised faster than the global average, i.e., >3.2 mm/year (IPCC, 2021).	Since the mid-1970s, droughts in the region have become more common, especially in the tropics and subtropics (IPCC, 2021).
East Asia	Overall, temperature showed an increasing trend of 0.255°C/decade over 1971–2014.	Summer Monsoon precipitation has showed a strong interdecadal variability over 1958–2008 (Zhang et al., 2013).	Rate of sea level rise over 1993–2016 was slower than the global mean estimated at 3.2 mm/year (Hens et al., 2018).	Over the past three decades, droughts have become more frequent in much of continental East Asia (IPCC, 2021).
West Asia	The region has warmed particularly in recent decades and this warming is more pronounced during winter than in other seasons (Alizadeh and Lin, 2021).	Overall, the region faced downward trend in precipitation over 1971–2014 (e.g., Tahroudi and Nejad, 2017).	Rate of sea level rise over 1993–2016 was about 9.6 mm/year (Hens et al., 2018).	Trend of warming has intensified droughts since the 1980s (IPCC, 2021).
Central Asia	Annual average temperature has increased by 0.32°C/decade during 1950–2016 (Haag et al., 2019).	Over 1950–2016, the annual precipitation increased by 3.13 mm/decade (Haag et al., 2019).	(-)	Heatwaves has increased by 1.3 times since the 1960s, with notable increasing trends in their intensity and duration.

4.3.1   CHANGE IN AIR TEMPERATURE

In recent decades, the Asian region has undergone significant changes in air temperature with far-reaching impacts on human health, ecosystems, and the economy (Portner et al., 2022). Several studies (e.g., Zhang et al., 2019; Ren et al., 2021) have investigated these changes and their findings indicate that temperature trends in Asia vary considerably across different regions (see Table 4.1).

- **Temperature trend in East Asia**
  East Asia has experienced an abnormal warming trend over the past decades, with an unprecedented increase in temperature during 1950–1998 (Ren et al., 2021). The overall temperature showed an increasing trend of 0.255°C/decade over 1971–2014. However, this warming trend varied by region; Northern areas of East Asia exhibited the highest warming trend, while Southeastern China experienced the lowest warming over 1901–2014. Additionally, the minimum temperature rose at a higher rate (0.33°C/10 years) than the maximum temperature (0.07°C/10 years) (Ren et al., 2021). This warming trend in East Asia was strongly influenced by high GHG emissions, particularly in China, and other anthropogenic factors (Xu et al., 2015). The region also experienced a similar warming trend in extreme temperature indices over the past four decades (Ren et al., 2021). For example, in China, the annual number of ice days decreased at a rate of approximately −2.3 days/10 years during the period 1961–2008 (Zhou and Ren, 2011; Ren et al., 2021).
- **Temperature trend in Central Asia**
  Central Asia has been identified as a "hotspot" for climate change (Giorgi, 2006). Over the past few decades, the temperature in Central Asia has risen significantly, particularly during winter seasons (Haag et al., 2019). This warming trend experienced an abrupt acceleration in intensity in the mid-1990s (Haag et al., 2019). Several studies have evaluated the past warming trend over Central Asia (e.g., Chen et al., 2009; Zhang et al., 2019; Haag et al., 2019). Although these studies confirmed a strong warming trend over the region, they reported different warming rates. For instance, Zhang et al. (2019) investigated changes in air temperature in the region over 1957–2005 and found that the annual average temperature rose at a rate of 0.32°C/decade during that period, indicating that the region was warming faster than the global average. They also found that the minimum temperature rose at a higher rate (0.41°C/decade) than the maximum temperature (0.24°C/decade). Another study by Haag et al. (2019) showed that the annual average temperature increased by 0.32°C/decade during 1950–2016. Furthermore, Chen et al. (2009) reported an annual temperature increase of 0.39°C/decade over 1979–2011. The different warming rates are likely attributable to variations in regional extents, time periods, and data sources (Haag et al., 2019).
- **Temperature trend in South Asia**
  South Asia, home to approximately one-fifth of the world's population, has experienced changes in air temperature over the past few decades that have emerged as a serious climate change issue (Lal et al., 2011). Temperature observations have revealed rising temperature trends over most countries

in South Asia, which are more pronounced during winter than in summer (Naveendrakumar et al., 2019). For example, in India, Sanjay et al. (2020) reported that the annual average temperature rose at a rate of 0.32°C/decade between 1986 and 2015. In Pakistan, several studies (e.g., Sheikh et al., 2009; Ahmad and Mahmood, 2017; Heureux et al., 2022) have reported that maximum and minimum temperatures increased by up to 1°C between 1960 and 2007, with a higher rise in Northern areas than in Southern areas.

- **Temperature trend in North Asia**
  Despite the scarcity of detailed temperature-specific historical data for North Asia, further research on past temperature changes in this region is imperative. An earlier study on temperature changes in Northern Asia from 1950 to 2005 and found an overall increase in winter and spring temperatures by 5 to 8°C/century. However, their findings also indicated that this warming trend was spatially nonuniform. The highest spatial variability of temperature trends was observed during the winter season, while summer and spring temperatures exhibited the lowest variations.
- **Temperature trend in West Asia**
  Previous studies have demonstrated a lack of exploration into warming trends in West Asia (e.g., Sivakumar et al., 2013). As such, further investigation is imperative. A recent study by Alizadeh and Lin (2021) examined temperature changes in West Asia from 1979 to 2018 and found that the region has experienced significant warming in recent decades, with a more pronounced warming trend during winter compared to other seasons.

4.3.2 CHANGE IN PRECIPITATION PATTERNS

In recent decades, the increase in GHG emissions and associated warming trends have resulted in more erratic precipitation patterns throughout Asia (Kim et al., 2019). The quantification of these changes is crucial due to their potential impact on human populations, infrastructure, and natural ecosystems (Kim et al., 2019). Numerous studies have investigated precipitation patterns in Asia (e.g., Kim et al., 2019; Supari et al., 2020; Muhammad et al., 2021; Ngai et al., 2022), often utilizing the Asian Precipitation-Highly-Resolved Observational Data Integration Towards the Evaluation of Water Resources (APHRODITE) project as a primary source of observational precipitation data due to its high-density rain gauge observations and extensive temporal coverage (Yatagai et al., 2012; Ngai et al., 2022). Key findings from these studies are summarized in Table 4.1.

- First, while precipitation trends have varied widely throughout the region over the past six decades, an overall decrease in annual rainfall of up to 15% has been observed for the entire Asian domain (Climate Centre, 2021). This variability is likely due to the redistribution of moisture from oceans to interior land as a result of increased temperatures (Deng et al., 2014).
- Second, precipitation in Eastern Asia is primarily concentrated during the East Asia Summer Monsoon, which has exhibited strong interdecadal variability in recent decades (e.g., Ding et al., 2008; Zhang et al., 2013).

- Third, in Central Asia, no significant long-term statistical trend in annual or seasonal precipitation was observed from 1950 to 2016 (Haag et al., 2019).
- Fourth, North and Central Asia experienced a significant upward trend in precipitation for all seasons except summer from 1971 to 2016 while South Asia saw a significant downward trend in winter precipitation over the same period.
- Finally, the West Asian region exhibited an overall downward trend in precipitation from 1971 to 2014 (e.g., Tahroudi and Nejad, 2017).

4.3.3   SEA LEVEL CHANGES

Sea-level rise poses a significant threat to coastal areas in Asia, where millions of people are at risk of flooding from coastal inundation and surge events (Brecht et al., 2012). Numerous studies have investigated past sea-level changes across the Asian domain (e.g., Hens et al., 2018; Swapna et al. 2020; Harrison et al., 2021). While these studies confirm a significant sea-level rise trend across the entire domain, exceeding the global average, they also reveal that sea-level rise has not been uniform across the region, resulting in varying rates of sea-level rise. These variations are likely due to local conditions such as ocean circulation, earthquakes, and variations in Earth's gravity field, as well as differences in time periods and methodologies (Hens et al., 2018). In West Asia, the rate of sea-level rise from 1993 to 2016 was approximately 9.6 mm/year, three times faster than the global average (Hens et al., 2018). In contrast, East Asia experienced a slower rate of sea-level rise than the global mean (Hens et al., 2018). The IPCC (2021) reports that sea-level rise around South Asia has exceeded the global average.

4.3.4   EXTREME CLIMATE EVENTS

Over the past few decades, particularly since the 1970s, the GHG emissions discussed in Section 2 have led to several extreme climate events across large regions of Asia. Table 4.1 provides a comprehensive list of observed extreme climate events for various regions of Asia. As indicated in Table 4.1, extreme climate events within the Asian domain primarily comprise intense and prolonged droughts, severe floods, heatwaves, and extreme precipitation. These events have inflicted significant damage on both the natural environment and human populations. For instance, according to the disaster database maintained by the Centre for Research on the Epidemiology of Disasters (CRED), an average of 36 flood events occurred annually in East Asia between 2000 and 2020, resulting in 3,166 fatalities. Additionally, extreme flood events in Pakistan in 2022 caused approximately $15 billion USD in economic damage (World Bank, 2022).

4.4   FUTURE CLIMATE TRENDS BY 2050

Recently, regional climate projections for the Asian domain have been generated using knowledge of various contributions applied to the outputs of coupled ocean-atmosphere climate model simulations, such as those produced through the sixth phase of the Coupled Model Intercomparison Project (CMIP6) (e.g., Almazroui et al., 2020b; Jiang et al., 2020). These studies have provided detailed information on future climate trends up to 2050 for different regions of Asia.

4.4.1 Projected East Asia Climate Change

For East Asia, it is projected that by 2050, temperature increases will be slightly lower than the global average compared to the present climate. Rainfall is expected to increase in northern regions (e.g., rainfall in Japan is projected to increase by approximately 87 mm) while decreasing in the Maritime Continent (IPCC, 2021; Climate Centre, 2021). Additionally, land subsidence and local human activities are anticipated to result in higher flood levels and prolonged inundation in the Mekong Delta (IPCC, 2021).

4.4.2 Projected South Asia Climate Change

For South Asia, projections for 2050 under the RCP 8.5 climate change scenario (i.e., high GHG emissions scenario) indicate that annual precipitation will increase by 12 mm in Pakistan to approximately 180 mm in the Maldives while a decrease of about 2.8 mm is projected for Afghanistan (Climate Centre, 2021). Almazroui et al. (2020b) found that, relative to the average climate over 1995–2014, the annual average temperature across the South Asian region is expected to rise by 0.9°C, 0.9°C, and 1.2°C under low (SSP1-2.6), medium (SSP2-4.5), and high (SSP5-8.5) GHG emissions scenarios, respectively. The same study also showed that these expected temperature changes are largest over the northwestern parts of India and Pakistan. Under all future climate scenarios, there are differences in projected sea level change across the South Asian domain. For instance, Harrison et al. (2021) revealed that the equatorial Indian Ocean will experience the highest sea-level rise, while the lowest rise will occur in the northern Arabian Sea and northern Bay of Bengal.

4.4.3 Projected Central Asia Climate Change

Previous studies (e.g., Jiang et al., 2020; Didovets et al., 2021; Vakulchuk et al., 2022) have investigated climate change in Central Asia and revealed that the region is already experiencing critical climate change, which is expected to intensify in the future. The projected change in temperature is relatively homogeneous across the region. Compared to the present day, average annual temperature by 2050 is expected to increase by 2.2°C in Turkmenistan to 2.7°C in Kazakhstan under the RCP 8.5 climate change scenario (Climate Centre, 2021). In contrast, the projected change in precipitation exhibits no clear trend or estimation (Didovets et al., 2021). Overall, Jiang et al. (2020) found an increase in annual average precipitation for the entire Central Asian region of approximately 4 to 14% under low and high GHG emission scenarios, respectively, by the end of the century (relative to the present-day). Additionally, an increase in extreme temperatures, frequency of droughts, and frequency, intensity, and duration of heatwaves is also anticipated for the region over the coming decades (e.g., Reyer et al., 2017; Feng et al., 2018; Vakulchuk et al., 2022).

4.4.4 Projected West Asia Climate Change

According to Sivakumar et al. (2013), the annual average temperature in West Asia is projected to increase between the present and the 2050s. This warming trend is anticipated to alter the distribution of extreme events, resulting in an increased frequency of heat waves. However, changes in precipitation patterns are less certain. While the subcontinental average winter precipitation is likely to increase, summer precipitation is expected to decrease in many parts of the region (Cruz et al., 2007).

4.4.5 Projected North Asia Climate Change

Cruz et al. (2007) projected that the most significant warming trend across the entire Asian domain in the coming decades will occur at high latitudes in North Asia. Furthermore, the subcontinental average winter precipitation is very likely to increase, while summer precipitation is also expected to increase.

4.5 ROLE OF ASIA IN ACHIEVING GLOBAL CLIMATE NEUTRALITY BY 2050

Nishioka (2016) predicted that if the Asia region continues its current trajectory toward highly energy-dependent societies, it will account for approximately half of the global economic power, energy consumption, and CO_2 emissions by 2050. This will make Asian countries major contributors to the global climate change issue. To mitigate this issue or at least minimize its impact on human and ecosystems, Asian countries must pursue a development path that promotes low GHG growth in the region. Achieving low GHG growth in the coming years is essential for transitioning to climate neutrality by 2050, as called for by the United Nations (UN) in 2021. The strategies required to achieve low GHG growth may vary significantly across the Asian domain due to factors such as geographical location, level of development, resource availability and type, climate conditions, and culture. However, the Low-Carbon Asia Research Project (2013) identified several common fundamental strategies for realizing low GHG growth in Asia (Figure 4.1).

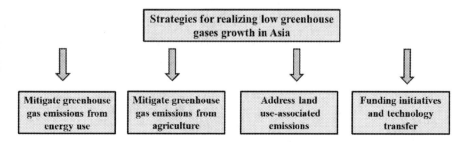

FIGURE 4.1 Strategies for realizing low greenhouse gases growth in Asia.

- Strategy 1: Promote actions and measures to mitigate CO_2 emissions from energy use and production. This includes enhancing energy efficiency, expanding production from domestic and renewable energy sources, and reducing reliance on fossil fuels (Fulton et al., 2017). With China, Japan, and the Republic of Korea moving away from foreign investments in coal, the region is closer to phasing out coal from the energy sector. However, a critical review of transition pathways in the Asia-Pacific region is necessary (UNESCAP, 2022).

- Strategy 2: Promote actions and measures to reduce CH_4 and N_2O emissions from the agriculture sector. A gradual transformation toward greening Asia's agricultural development is essential. Various agricultural measures have been proposed and tested for achieving green agriculture in Asia, such as reducing CH_4 and N_2O emissions through conservation and climate smart agriculture, cultivating new crop varieties adapted to climate change, implementing green farming technologies, improving livestock management, preventing soil salinity and erosion, and applying agroecological measures (Aryal et al., 2020). However, institutional support for implementing and disseminating these technical measures must be strengthened (Aryal et al., 2020).

- Strategy 3: Address land use-associated emissions and CO_2 absorption. This can be achieved by optimizing land use structure, controlling land use intensity rationally, and promoting efficient use of urban construction land through scientific spatial land-use planning and sensible land-use policies (Tang and Hu, 2021).

- Strategy 4: Ensure effective implementation of the above strategies through good governance, funding initiatives, and technology transfer. Using technology is a common strategy for adapting to climate change (Nor Diana et al., 2022). Enhancing financial flows consistent with a pathway toward low GHG emissions is also crucial.

The Low-Carbon Asia Research Project (2013) suggests that if all the strategies are effectively implemented, GHG emissions in Asia can be significantly reduced by 2050. However, effective implementation of these strategies must account for variations in geography, climate conditions, level of development, and resource availability and type across the Asian region (Azhgaliyeva and Rahut, 2022). Long-term planning and continuous learning from other countries, particularly European countries, will also be necessary (Azhgaliyeva and Rahut, 2022). Additionally, developing Asian countries will require substantial support to meet their nationally determined contributions, improve energy efficiency, and reduce the impact of climate disasters (Azhgaliyeva and Rahut, 2022).

4.6 CONCLUSIONS

Asia, led by China, is one of the largest contributors to global climate change due to its large population, rapid economic growth, and high energy consumption. This chapter demonstrated that the region, particularly China, has been responsible for over 50% of global GHG emissions in recent decades. As a direct consequence of these emissions, since the 1970s, all Asian countries have experienced various forms of climate change, including warming trends, droughts, changes in precipitation patterns, and rising sea levels. Climate projections suggest that without adequate mitigation measures, these impacts will worsen in the coming decades. Therefore, it is imperative that the Asian community prepares for this critical situation through appropriate adaptation strategies. Strategies such as promoting renewable energy sources, green agriculture, and optimizing land use can support the region's transition to net-zero GHG emissions by 2050. Effective implementation of these strategies would enhance the region's ability to address the global climate change issue.

EXERCISE (WITH ANSWERS)

Question: What are the main information sources for exploring the climate change issue in Asia?

Answer: There are several sources that provide valuable data and information on the climate change issue in Asia. Some examples include:

1. The Intergovernmental Panel on Climate Change (IPCC): This international body provides comprehensive assessments of the science, impacts, and policy responses related to climate change. Its reports offer a wealth of information on climate change projections and impacts in Asia and other regions (https://www.ipcc.ch/).
2. The Central Asian Climate Information Platform (CACIP): This platform provides useful information on climate adaptation and mitigation in Central Asia (https://ca-climate.org).
3. The Regional Climate Consortium for Asia and the Pacific (RCCAP): This consortium provides climate information for many Asian countries, including historical climate observations and future climate trends under different GHG emissions scenarios. Its portal can be used to identify and access climate information relevant to adaptation planning needs (https://www.rccap.org/about/).
4. The Asia-Pacific Climate Change Adaptation Information Platform (AP-PLAT): This online platform bridges science and stakeholders to support climate change adaptation in the Asia-Pacific region (https://ap-plat.nies.go.jp/).
5. The Southeast Asian Climate Assessment & Dataset (SACA&D): This resource combines the collation of daily series of observations at meteorological stations with quality control, analysis of extremes, and dissemination of both the daily data and analysis results (https://www.climateurope.eu/southeast-asian-climate-assessment-dataset-sacad/).

5 Climate Change in North America

Past Indices, Future Trends, and Pathways to Net-Zero Greenhouse Gas Emissions by 2050

5.1 INTRODUCTION

North America has experienced significant and rapid climate change in recent decades, characterized by rising air temperatures (Gil-Alana and Sauci, 2019), altered precipitation patterns (Vincent et al., 2018), and an increased frequency of extreme weather events such as droughts (Overpeck and Udall, 2020), heatwaves (Monier and Gao, 2015), and floods (Tate et al., 2021). These changes are primarily attributed to the substantial increase in greenhouse gas (GHG) emissions, particularly carbon dioxide (CO_2), methane (CH_4), and nitrous oxide (N_2O), over the past century and their accumulation in the atmosphere (IPCC, 2021). North America, led by the United States, is the second-largest emitter of GHGs globally, following China (Ge et al., 2020). From 1990 to 2019, the United States alone accounted for approximately 12.5% of global GHG emissions (Ge et al., 2020). Canada is also among the top ten GHG emitters worldwide, with emissions of approximately 736 million metric tons (Mt) of CO_2 equivalent (CO_{2e}) in 2019 (Ge et al., 2020). The primary sources of GHG emissions in North America are energy production and consumption and the agriculture, forestry, and other land use (AFOLU) sector (Lamb et al., 2021). Recent climate projections for the year 2050 indicate an extensive warming trend in most areas of North America, changing precipitation patterns, increases in maximum precipitation levels, and a significant acceleration of sea level rise as a direct result of increasing atmospheric concentrations of GHGs (Asong et al., 2019; Singh et al., 2022; Sweet et al., 2022).

The anticipated climate change, through direct and indirect means, has the potential to impact all sectors of the economy, damage infrastructure, and result in significant human losses (Monier and Gao, 2015). In response, North American countries, including the United States and Canada, have developed long-term plans to decarbonize their economies and transition to net-zero GHG emissions by 2050 (Busch et al., 2023). Effective decarbonization plans can be identified and developed based on historical GHG emissions data, future climate information under various GHG emissions scenarios, and future climate targets. Policymakers require these factors

DOI: 10.1201/9781003404194-6

to develop strategies and policies for emissions reductions and to monitor progress toward achieving agreed-upon climate targets by 2050 (Busch et al., 2023). This chapter aims to address three questions: (1) What role does North America play in global GHG emissions? (2) What will the future climate be like in the region in 2050? and (3) What pathways has the region adopted to reach net-zero GHG emissions by 2050? The focus is on the United States and Canada as they are the largest GHG emitters in North America. The findings of this chapter should provide readers with valuable information about climate change issues in both countries.

5.2 HISTORICAL GREENHOUSE GAS EMISSIONS

5.2.1 USA Greenhouse Gas Emissions

In 2021, the United States emitted approximately five billion metric tons of CO_{2e} in GHG emissions, making it the second-largest GHG emitter globally, following China (Ritchie et al., 2020). Carbon dioxide emissions, primarily from the combustion of fossil fuels for energy production and industrial activities such as cement production, accounted for approximately 79% of total USA GHG emissions (Ritchie et al., 2020). Methane and N_2O emissions from agricultural activities such as synthetic fertilizer use and rice cultivation also contributed to USA GHG emissions (Ritchie et al., 2020). However, these emissions have received less attention from researchers compared to CO_2 emissions (Tian et al., 2015). USA GHG emissions increased between 1900 and 2010 before decreasing significantly over the past decade (Ramseur, 2014). This reduction can be partially attributed to a shift in electricity production sources, with a 100% increase in electricity generation from renewable sources such as wind, hydroelectric, and solar power between 2004 and 2013 (Ramseur, 2014). The COVID-19 pandemic lockdown measures also played a role in reducing GHG emissions in 2020 (Ray et al., 2022).

Factors that can affect GHG emission levels include energy intensity, population growth, and electricity production methods (Ramseur, 2014) (Figure 5.1). A study by Khan and Liu (2023) found that promoting renewable energy sources for electricity production and reducing energy intensity can significantly decrease CO_2 emissions and promote a cleaner environment. This is particularly important given the USA' pledge to reduce its GHG emissions by approximately 50% below 2005 levels by 2030 (den Elzen et al., 2022).

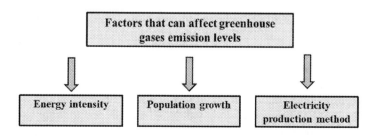

FIGURE 5.1   Factors that can affect greenhouse gases emission levels.

5.2.2 CANADIAN GREENHOUSE GAS EMISSIONS

Between 1990 and 2007, factors such as population growth, increased affluence, and continued reliance on fossil fuels led to a 21% increase in GHG emissions in Canada (Kerr and Mellon, 2012). Approximately 80% of these emissions were in the form of CO_2 (Kerr and Mellon, 2012). However, other potent GHGs such as CH_4 and N_2O also contributed to climate change in Canada (Ritchie et al., 2020). CH_4 emissions primarily originated from fugitive emissions from oil and natural gas systems, coal mining, agriculture and animal waste management systems, landfills, and wastewater treatment (Fouli et al., 2021). N_2O emissions were mainly from agricultural soil management, energy and fuel combustion, industrial processes, and waste management (Fouli et al., 2021). Canada's failure to meet its international commitments, such as the Kyoto Protocol, which required the federal government to reduce GHG emissions to 6% below 1990 levels by 2012, also contributed to the increase in GHG emissions between 1990 and 2007 (Kerr and Mellon, 2012). However, GHG emissions varied significantly among cities due to differences in primary energy sources, climate conditions, and population densities (Welegedara et al., 2021).

Between 2015 and 2021, a shift from an increase to a decrease in GHG emissions was observed in Canada (Nascimento et al., 2022). In 2018, total GHG emissions in Canada amounted to only 729 megatons of CO_{2e} (ECCC, 2018). This shift can be attributed to factors such as the transition to renewable energy sources, the promotion of sustainable agriculture practices, and the implementation of a carbon tax strategy. These measures have the potential to help Canada reduce its GHG emissions by approximately 45% below 2005 levels by 2030.

5.3 PAST AND FUTURE TRENDS OF CLIMATE CHANGE

5.3.1 CLIMATE CHANGE IN THE UNITED STATES

Numerous studies have examined both historical and projected changes in climate variables across the United States (e.g., Ganguli and Ganguly, 2016; Vose et al., 2017; Easterling et al., 2017; Sweet et al., 2022). These investigations have yielded several key findings, as illustrated in Figure 5.2.

- According to Vose et al. (2017), the annual average temperature across the United States increased by 0.7°C between 1986 and 2016 when compared to the reference period of 1901–1960. However, it is important to note that the rate of warming varied significantly among regions. For instance, Alaska experienced the greatest warming (rate >0.6°C) while the Southeast region experienced the least warming (<0.3°C). Overall, the warming trend in the country has accelerated significantly since 1979 (Vose et al., 2017). Future climate projections indicate that this warming trend is expected to continue through 2023–2050 under all GHG emissions scenarios (i.e., all RCP scenarios). Additionally, extreme temperatures in the country are also projected to increase (Vose et al., 2017).
- Scientists have also observed significant changes in precipitation patterns over recent decades. On average, annual precipitation has increased by

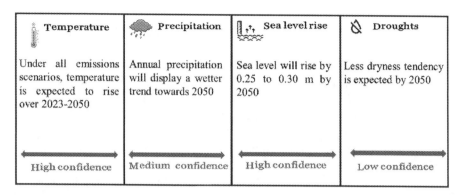

Temperature	Precipitation	Sea level rise	Droughts
Under all emissions scenarios, temperature is expected to rise over 2023-2050	Annual precipitation will display a wetter trend towards 2050	Sea level will rise by 0.25 to 0.30 m by 2050	Less dryness tendency is expected by 2050
High confidence	Medium confidence	High confidence	Low confidence

FIGURE 5.2 Climate change projections for the USA by 2050.

approximately 4% since 1900, particularly in the Northern and Southern regions (Easterling et al., 2017). However, some regions, mainly in the Western, Southwestern, and Southeastern USA, have become drier (Easterling et al., 2017). Furthermore, since that year, the country has experienced more frequent heavy precipitation events (Easterling et al., 2017). Compared to the climate of the period from 1995 to 2014, the projected annual precipitation is expected to display a wetter trend toward 2050 under different RCP scenarios. By 2100, precipitation is expected to increase by 10%–30% across the country under the SSP5-8.5 scenario. Future climate projections also suggest that the warming trend in the United States is expected to further increase precipitation intensity as well as the magnitude and frequency of precipitation events during the upcoming decades (Easterling et al., 2017).

- In the United States, sea level rise is considered a strong indicator of ongoing climate change (Sweet et al., 2017). Past tide gauge observations have shown that the average sea level in the country has risen significantly over recent decades (Sweet et al., 2022). However, the rate of rise varies greatly along the coastlines of the United States due to factors such as changes in Earth's gravitational field and rotation from melting land ice, changes in ocean circulation, and vertical land motion (Sweet et al., 2017). On average, the sea level in the United States has increased by 6.5 inches since 1950, with about half of this rise occurring in the past 20 years (https://sealevelrise.org/). Scientists predict that the rising trend in sea level in the United States will continue at an accelerated rate through 2050. Compared to the rise measured between 1920 and 2020, the sea level in the country is expected to rise by an additional 0.25 to 0.30 meters between 2023 and 2050 (Sweet et al., 2022). This is expected to result in more frequent coastal flooding events. As tens of millions of people in the United States already live in coastal areas, future sea level rise and associated flooding events pose significant risks to both people and infrastructure.

- Drought is a recurring issue in the United States. Since 1980, the country has experienced several catastrophic droughts resulting in significant economic damages estimated at approximately $210.1 billion between 1980 and 2011 (Smith and Katz, 2013). Long-term analyses of drought trends over recent decades have shown no significant trend in average drought severity across the entire country (e.g., Pal et al., 2013). However, regional analyses have revealed increasing trends in extreme drought severity (e.g., Ford and Labosier, 2014; Ganguli and Ganguly, 2016). Droughts in the United States can also exhibit temporal trends. For example, in 2012, approximately 55% of the total land area of the United States was classified as experiencing moderate drought while in 2017, only 3.8% of the area was affected by drought of at least moderate intensity (Folger, 2017). Estimating future trends in drought (including trends in duration, intensity, and severity) is challenging and complex due to the high variability in atmospheric and soil moisture on small spatial scales (Gamelin et al., 2022). With low to medium confidence, the 2012 report of the Intergovernmental Panel on Climate Change (IPCC) indicates a slight overall tendency toward less dryness in the United States through 2050. However, other studies (Cook et al., 2015; Duffy et al., 2015; Gamelin et al., 2022) suggest that under global warming, drought extremes are expected to increase throughout the 21st century, ultimately affecting water resources, wildfire activity, and crop loss.

5.3.2 CLIMATE CHANGE IN CANADA

Canada, like several other regions across the globe, has experienced significant climate change over the past few decades. This phenomenon has had profound impacts on the country's environment, economy, and public health (Warren and Lemmen, 2014). Numerous scientific studies have documented past trends and projected future changes in climate (e.g., Asong et al., 2018; Li et al., 2018; Zhang et al., 2019; Greenan et al., 2019), providing decision makers with robust climate information to inform policy approaches to climate change and natural hazards (Newton et al., 2005). The key findings of these studies are summarized in Figure 5.3.

- Long-term analyses of past and projected future changes in Canada's temperatures have revealed several key findings. First, between 1948 and 2016, the annual average temperature across the country increased by an average of 1.7°C. However, in some regions, particularly Northern Canada, this

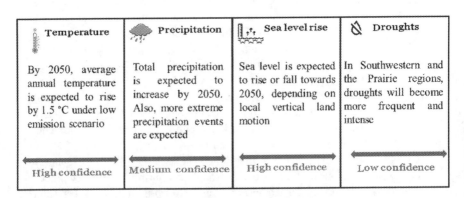

FIGURE 5.3   Climate change projections for Canada by 2050.

increase reached 2.3°C (Vincent et al., 2015; Zhang et al., 2019), indicating that Canada is warming at twice the rate of the global average (Zhang et al., 2019). Second, warming in Canada varies significantly by season. Zhang et al. (2019) reported that from 1948 to 2016, the average temperature increased by 3.3°C in winter and only by 1.5°C in summer. Third, future simulations predict that Canada will continue to warm in the coming decades, regardless of GHG emissions scenarios. On average, compared to the recent past period (1986–2005), the average annual temperature in Canada by 2050 is projected to increase by 1.5°C under a low GHG emissions scenario (RCP2.6) and by 2.3°C under a high emissions scenario (RCP8.5) (Zhang et al., 2019). However, it is important to note that future temperature projections vary significantly by region and season (Zhang et al., 2019). For example, under the RCP2.6 scenario, temperatures by 2050 are expected to increase by 1.8°C in the North, while in British Columbia they will rise only by 1.3°C (Zhang et al., 2019). Finally, relative to the period from 1986 to 2005, a significant increase in cooling degree-days for Southern regions of the country and a significant reduction in heating degree-days for Northern regions are expected toward 2050 (Li et al., 2018).

- On average, precipitation in Canada has increased by 20% from 1948 to 2012 (Vincent et al., 2015; Zhang et al., 2019). By 2050, scientists expect an overall increase in total precipitation, frequency and intensity of precipitation, and extreme precipitation across the entire Canadian domain with medium confidence (e.g., Li et al., 2018; Zhang et al., 2019). However, for some locations (e.g., some areas in Atlantic Canada), a small decrease in the amount of precipitation is expected (Li et al., 2018). The country should also anticipate an increase in the number of days with precipitation in Northern regions and a small reduction in some areas of the South (Li et al., 2018).

- Since much of Canada's landmass is composed of glaciers, ice sheets, and snow, the warming trend in the country can lead to melting of these components. This facilitates their runoff into seas, causing sea levels to rise at a dramatic rate. Sea level rise along Canada's coastline varies significantly by region. For instance, along the southeast Atlantic coast, sea level has risen at an average rate of 3 mm per year over the past 50 years while along the Pacific and Arctic coasts, it has risen at a rate of 1.7 mm per year. This indicates that trends in regional average sea levels can differ significantly from the trend in global average. Relative to the period from 1986 to 2005, sea level in Canada is expected to rise or fall toward 2050 depending on local vertical land motion (Greenan et al., 2019). Specifically, it is expected to rise along the Atlantic and Pacific coasts (Greenan et al., 2019), which may lead to an increase in the frequency and magnitude of extreme coastal flooding events (Greenan et al., 2019), while it is projected to fall in regions where land is uplifting relatively quickly such as the Hudson's Bay region (Greenan et al., 2019).

- Research on drought properties is crucial for Canadian decision makers, as drought is among the most costly natural hazards (Bonsal et al., 2011; Mardian, 2022). Using various climatological drought indicators, particularly the Standardized Precipitation Evapotranspiration Index (SPEI), researchers in Canada have identified several periodic droughts over the past few decades with varying durations and severities, such as devastating drought events across Western Canada from 1999 to 2005 and in 2015 (Asong et al., 2018). Between 1950 and 2013, drought events predominantly affected the Prairie region due to its location in the lee of the Rocky Mountains and its high dependence on rainfed agriculture (Asong et al., 2018). Drought projections in Canada based on the SPEI from the Coupled Model Intercomparison Project 5 (CMIP5) indicate that droughts in Southwestern Canada and the Prairie region are expected to increase in frequency and intensity (Tam et al., 2018), potentially affecting sensitive economic sectors such as agriculture. This underscores the need for enhanced adaptation measures.

5.4 PATHWAYS TO NET-ZERO GHG EMISSIONS BY 2050

5.4.1 USA's Pathways to Net-Zero Emissions

In the United States, achieving the 2050 net-zero emissions goal requires reducing net GHG emissions from approximately 5.7 Gt CO_{2e} in 2020 to zero by no later than 2050 (United States Department of State and the United States Executive Office of the President, 2021). To accomplish this, the United States launched a long-term strategy in 2021 for achieving its net-zero emissions target by 2050. The strategy is based on key transformations (Figure 5.4)

- Transformation 1: Electricity Decarbonization
  Electricity decarbonization is a central focus of climate policy in the United States and represents a crucial step toward achieving a low-carbon future. This is due to its potential to reduce the environmental impact of all sectors (Kiss and Szalay, 2022). In recent years, the United States has experienced an accelerated transition toward a clean electricity system. This transition has been driven by factors such as decreasing costs for solar and wind technologies, federal and subnational policies, and consumer demand

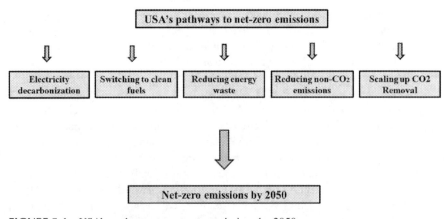

FIGURE 5.4 USA's pathways to net-zero emissions by 2050.

(United States Department of State and the United States Executive Office of the President, 2021). However, further efforts are necessary to achieve deeper electricity decarbonization in the coming years. This will build upon current successes and enable the country to reach its goal of 100% clean electricity by 2035, a critical component of achieving net-zero emissions by 2050 (United States Department of State and the United States Executive Office of the President, 2021). To promote deeper electricity decarbonization, an encompassing policy package that combines and reinforces energy and climate policy incentives is required (Bang, 2021).

- Transformation 2: Switching to Clean Fuels
  The primary goal of this transformation is to improve energy efficiency and reduce GHG emissions from industrial processes by transitioning from more emissions-intensive fossil fuels, such as oil, to electricity in various sectors, including transportation and buildings (United States Department of State and the United States Executive Office of the President, 2021). In sectors where electrification presents technological challenges, such as aviation, shipping, and certain industrial processes, the use of clean fuels like carbon-free hydrogen and sustainable biofuels can be prioritized (United States Department of State and the United States Executive Office of the President, 2021).

- Transformation 3: Reducing Energy Waste
  Clean energy technologies represent a promising solution for reducing GHG emissions. The transition to cleaner sources of energy can be accelerated and made more cost-effective by improving the energy efficiency of both existing and new technologies. This allows for the provision of the same or better services while using less energy (United States Department of State and the United States Executive Office of the President, 2021). In recent years, investment in clean energy technologies, companies, and projects has increased in the United States. Continued investment is necessary to achieve the country's goal of net-zero emissions by 2050 (United States Department of State and the United States Executive Office of the President, 2021).

- Transformation 4: Reducing Non-CO_2 Emissions
  Non-CO_2 gases, such as CH_4 and N_2O, are significant contributors to global warming. For example, CH_4 alone is responsible for half of the current net global warming of 1.0°C (United States Department of State and the United States Executive Office of the President, 2021). In response to this challenge, the United States is committed to implementing comprehensive and urgent measures to reduce CH_4 emissions. Through the Global Methane Pledge initiative, the United States and its partners aim to decrease global CH_4 emissions by at least 30% by 2030. This would prevent over 0.2°C of warming by 2050. The United States is also supporting research initiatives that can help unlock the innovation necessary for achieving deep emissions reductions beyond what is currently possible with available technologies (United States Department of State and the United States Executive Office of the President, 2021).

- Transformation 5: Scaling Up CO_2 Removal
  In the coming decades, GHG emissions from energy production can be progressively reduced until net-zero emissions are achieved in 2050. However, certain emissions, particularly non-CO_2 emissions from agriculture, will be difficult to completely decarbonize by this date. To reach net-zero CO2 emissions, it will be necessary to remove CO2 from the atmosphere using rigorously validated processes and technologies. This will require scaling up both land-based carbon sinks and engineered strategies (United States Department of State and the United States Executive Office of the President, 2021).

5.4.2 Canada's Pathways to Net-Zero Emissions

According to the IPCC (2022), there is still time to limit average global warming to 1.5°C above pre-industrial levels. However, it is crucial to take immediate action to reduce GHG emissions in the coming years. GHG emissions must be progressively reduced until net-zero emissions are achieved by 2050 in order to avoid the worst outcomes. To accomplish this goal, each country must develop appropriate tools, measures, legislation, and transformative policy choices that can facilitate the transition to net-zero emissions by 2050. In Canada, this task has been assigned to the Canadian Climate Institute (Canadian Climate Institute, 2021). The country has committed to reducing its GHG emissions to net zero by 2050 and has developed a range of policies, new legislation, and measures to reduce emissions and remove them from the atmosphere (Canadian Climate Institute, 2021). Previous studies have shown that achieving net-zero emissions by 2050 is possible for Canada through several pathways. Individuals with appropriate knowledge and skills can play a key role in this effort. The following are examples of these pathways (Figure 5.5):

1. Aligning Energy Production and Use with "Net-Zero Canada"
   In Canada, the energy sector is central to achieving the goal of net-zero emissions (Canadian Climate Institute, 2021). This sector has a key role to play and a significant responsibility to act toward this goal. Decision makers are carefully considering the need to develop sustainable, carbon-neutral energy in every region of Canada. This can be achieved by improving energy efficiency, reducing CO_2 emissions from industrial processes, transitioning from more emissions-intensive fossil fuels like oil to natural gas, and promoting the production and use of renewable and sustainable energy sources in various economic sectors, particularly transportation. Canada is considered a world leader in the production and use of energy from renewable resources, including moving water, wind, biomass, solar, geothermal, and ocean energy. These sources currently provide about 19% of Canada's total primary energy supply (Natural Resources Canada, 2017), and this percentage is expected to increase in the coming years (Natural Resources Canada, 2017). The production of energy from photoautotrophic microorganisms has emerged as a promising measure for balancing food, energy, and environmental needs due to their high CO_2 capturing efficiency (Maurya et al., 2021).

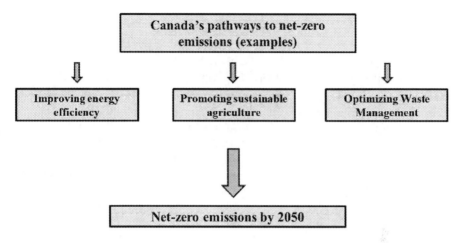

FIGURE 5.5   Canada's pathways to net-zero emissions (examples).

2. Promoting Agricultural Activities in Alignment with "Net-Zero Canada"
   It is important to note that achieving net-zero emissions in the agriculture sector is challenging due to a lack of abatement options (Navius Research, 2021). Agricultural GHG emissions have many sources, including enteric fermentation in cattle, the application of synthetic fertilizers, biomass decomposition, soil cultivation and tillage, and mineralization of soil organic matter and manure (Fouli et al., 2021). However, there are several options for reducing GHG emissions from this sector, depending on the type of agricultural activity (Fouli et al., 2021). Major options include conservation agriculture, agroforestry, restoring degraded land, increasing soil carbon content, forestry sequestration, and no-till or reduced tillage practices. Soil carbon content can be increased and stored in the soil or plants to reduce CO_2 losses to the atmosphere (Fouli et al., 2021). Carbon storage can be achieved by using agroforestry, cover crops, or mulches and switching from annual to perennial cropping (Fouli et al., 2021). No-till practices can improve soil properties such as porosity, moisture retention, and organic matter content, creating a healthy environment for roots, microorganisms, and fungi (Fouli et al., 2021). These options will play a crucial role in decarbonizing Canada's economy (Navius Research, 2021). Additionally, consumers can help reduce agricultural GHG emissions by making informed choices when purchasing food (Fouli et al., 2021).
3. Optimizing Waste Management
   Canada is among the countries that strongly encourage initiatives and options aimed at reducing GHG emissions from waste management. These include waste recycling, reducing landfill methane emissions, improving landfill gas collection efficiency, and promoting a circular economy (i.e., an economy that aims to reduce waste from production and distribution processes as one of its components) (Yang et al., 2023). To achieve the net-zero emissions target in the waste sector, it is crucial to increase public awareness about municipal solid waste management and its importance for effective waste management.

5.5 CONCLUSIONS

In this chapter, we reviewed the issue of climate change in North America, with a focus on the United States and Canada, based on numerous published papers. The chapter revealed that the United States has been a major contributor to GHG emissions in recent decades. As a result, the North American region has experienced various forms of climate change, including continued warming temperatures, droughts, changes in precipitation patterns, and rising sea levels. These changes may have significant impacts on the people of North America. Climate projections suggest that the situation may worsen by 2050, highlighting the need for North American countries to implement suitable adaptation strategies to avoid this critical situation. The United States has recognized its responsibility for the current global climate change issue and is committed to taking comprehensive and urgent measures to sharply reduce GHG emissions until net-zero emissions are achieved by 2050. Failure to do so could have catastrophic consequences for human civilization. In this context, further research initiatives are needed to unlock the innovation required for deep GHG emissions reductions.

Part II

Climate Change Effects
on Agriculture by 2050

Climate change is known to have various effects on agricultural productivity, and is projected that continued greenhouse gases emissions will lead to further effects over the coming few decades. As identifying past and future effects of climate change on agricultural productivity is highly needed to develop adequate measures and strategies that can simultaneously increase adaptive capacity of the agriculture sector and reduce greenhouse gases emissions, it is hoped that the second part of this book will help to provide decision makers, specialists, farmers, students, and nonspecialists comprehensive global, regional, and country data related to past as well as future (2050) effects of climate change on agricultural productivity.

6 Global Agricultural Production Under Climate Change by 2050
Implications for Global Food Security

6.1 INTRODUCTION

Previous studies (e.g., Papalexiou and Montanari, 2019; Yerlikaya et al., 2020) have confirmed that rising atmospheric concentrations of greenhouse gases (GHGs) will have significant impacts on Earth's climate in the coming decades. Even under optimistic scenarios, it is projected that the frequency of extreme climate events such as droughts and heatwaves will increase, sea levels will rise, ocean currents may reverse, and precipitation patterns will change (Zhai and Zhuang, 2009; IPCC, 2022). These various aspects of climate change could critically affect yields of major crops such as wheat, maize, and rice, with serious implications for global food security (Campbell, 2022; Farooq et al., 2023).

Climate remains a key factor in determining agricultural productivity worldwide, and predicted changes in climate over the coming decades and their associated effects on water availability, pests, and diseases are expected to substantially affect agricultural production potential (Zhai and Zhuang, 2009; IPCC, 2022). In a global analysis of 155 regions, Springmann et al. (2016) projected that global food availability would decrease by 3.2% by 2050 (relative to 2010) if global warming and its effects on agricultural production continue unabated, with many regions in developing countries potentially facing larger reductions. At the same time, due to population growth (from approximately 6.9 billion in 2010 to 9.4–10.1 billion by 2050), economic growth, and changing lifestyles, particularly in developing countries (United Nations, 2019), global demand for crops and grass is projected to increase from 5.5 to 10.9 Gt between 2007 and 2050 (Bijl et al., 2017). More recently, van Dijk et al. (2021) reviewed 57 global food security projection studies and found that global food demand is expected to rise by up to 56% between 2010 and 2050. To meet this future demand, global agricultural production must increase by up to 56% over the same period (Figure 6.1). Ensuring global food security for a projected population of 9.4 to 10.1 billion by 2050 will be a significant challenge for the international community and will increase the difficulty of managing food risk. As such, there is an urgent need to assess the potential impacts of climate change on global agricultural production by 2050, particularly in major producing countries (Teng et al., 2015).

DOI: 10.1201/9781003404194-8

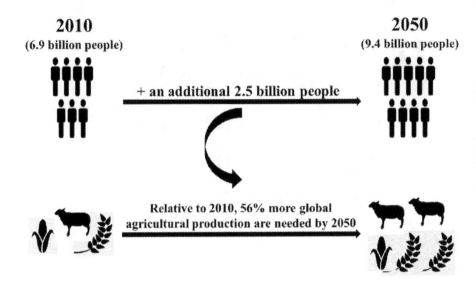

FIGURE 6.1 An increase (56%) in global agricultural production is required by 2050 to meet the food demands of an additional 2.5 billion population relative to 2010 levels. (Data sources: United Nations, 2019; van Dijk et al., 2021.)

To evaluate long-term global agricultural production under various socio-economic and climate change scenarios, researchers have employed advanced climate and agricultural models (e.g., van Dijk et al., 2021). Climate change scenarios are typically based on assumptions about future atmospheric carbon dioxide (CO_2) concentrations, which depend on changes in population, energy use, and land use (Cheng et al., 2022). However, due to these assumptions and other factors such as differences in the crops investigated and the year of projection, comprehensive analyses of global agricultural production projections are limited (e.g., Ahmed, 2022; Farooq et al., 2023). Therefore, the primary objective of this chapter is to provide a comprehensive analysis of global agricultural production projections by 2050. Decision-makers, leaders, farmers, and scientists worldwide require these projections to develop appropriate agricultural measures for the coming decades (IPCC, 2022). The specific objectives addressed in this chapter are (1) to review the methods and models used in global food production projections; (2) to review future global agricultural production under climate change by 2050 and compare it to food demand by that year (i.e., to assess food security by 2050); and (3) to identify countries that could play a key role in ensuring global food security by 2050.

6.2   MODELING OF CLIMATE CHANGE EFFECTS ON AGRICULTURAL PRODUCTION

Projections of the effects of climate change on agricultural production are primarily based on crop modeling (Soussana et al., 2010). However, modeling the effects of climate change on agricultural productivity (i.e., specifically, on crop growth and yield)

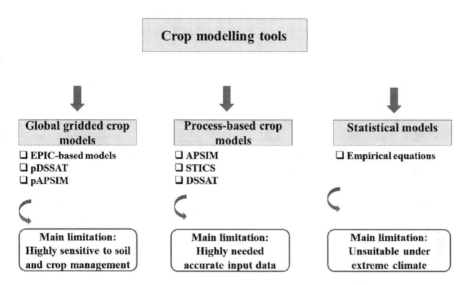

FIGURE 6.2   Main crop modeling tools.

is a complex task, and researchers worldwide employ a variety of methods, models, and tools to do so (Soussana et al., 2010). The main methods and models used are listed below (Figure 6.2).

6.2.1   GLOBAL GRIDDED CROP MODELS (GGCMS)

Crop models are useful for assessing the interactions between crops and agronomic management across space and time under various climate change scenarios (Pasley et al., 2023). Researchers worldwide consider crop models to be valuable tools for projecting future crop production under climate change and for making decisions and recommendations (Pasley et al., 2023). To assess the future effects of climate change on crop production, engineers typically link climate projections to crop models and land management decision tools (Tubiello et al., 2002; Soussana et al., 2010). In recent decades, many global gridded crop models (GGCMs) have been developed to help researchers predict crop yield and growth under different climate change scenarios (Camargo-Alvarez et al., 2022). GGCMs (e.g., EPIC-based models, pDSSAT, pAPSIM) are primarily used to simulate agricultural production, predict the effects of climate change on crop yield, and evaluate the utility of various management practices such as irrigation and fertilization (e.g., Franke et al., 2020; Ringeval et al. 2021; Camargo-Alvarez et al., 2022). Additionally, when combined with economic modeling, they can provide data for projecting future land use change (Müller et al., 2017). For more information about these models, see the IMPACT models (https:// www.isimip.org/impactmodels/). GGCMs have demonstrated good performance in simulating future global yields for major crops such as wheat and maize (e.g., Li et al., 2022). However, it is important to note that GGCMs are subject to uncertainties arising from model structure and parameterization as well as from calibration and input data quality (Müller et al., 2017). Therefore, it is essential to continue advancing our fundamental understanding of GGCMs, reducing uncertainties, and assessing future risks in order to obtain more accurate simulation results.

6.2.2 PROCESS-BASED CROP MODELS (PBMS)

Process-based crop models (PBMs) simulate crop growth in interaction with the environment using computer language, enhance our understanding of underlying processes, and enable scenario analyses (Seidel et al., 2022). PBMs integrate crop properties, climate, and environmental conditions in a systems approach to predict crop growth and yield under various scenarios at a local scale (IFPRI, 2009). They are also useful for quantifying carbon budgets at large scales (Brilli et al., 2017). Furthermore, PBMs are powerful tools for obtaining information about how a crop grows in interaction with its environment by identifying which crop properties are sensitive to model simulation conditions, how the outcomes of these properties respond to anticipated changes in growth conditions, and where high crop growth performance can be achieved (Weih et al., 2022). However, substantial time and data (especially from field experiments) are required to calibrate and validate these models for a given area (Wallach et al., 2006; Soussana et al., 2010), which can make their use challenging (Weih et al., 2022). Additionally, relevant climate data and soil and crop properties and mechanisms driving crop responses to climate change must be incorporated into these models. In the context of climate change studies, many PBMs have been developed in recent decades and are increasingly being applied to assess the effects of climate variability on crop growth and yield (Tao et al., 2009). Specifically, most PBMs have the ability to predict key crop responses to climate change such as CO_2 fertilization effects and changes in transpiration using a pro-portionality factor (Long et al., 2006). The most frequently used PBMs include the Agricultural Production Systems Simulator (APSIM) (Keating et al., 2003), the Decision Support System for Agrotechnology Transfer (DSSAT) (Hoogenboom et al., 2019), the Multidisciplinary Simulator for Standard Crops (STICS) (Brisson et al., 2003), and Daisy (Abrahamsen and Hansen, 2000).

6.2.3 STATISTICAL MODELS

Statistical crop models, also known as empirical models, are increasingly being used to assess crop yield and growth responses to climate change (e.g., Lobell and Burke, 2010; Holzkämper et al., 2012). These models establish relationships between crop yield and changes in climate variables (e.g., temperature and rainfall) through cor-relation analysis (Shem Juma and Kelonye Beru, 2021). The development of these models involves constructing empirical equations that describe crop yields based on a range of assumptions, which can result in modeling errors (Yang et al., 2014). Therefore, it is essential to evaluate the performance of these models against observed data (i.e., observed crop yields). Statistical assessment is considered a key method for comparing the outputs of statistical crop models with observed data (Yang et al., 2014). Metrics such as the Root Mean Squared Error (RMSE), normalized Root Mean Squared Error (nRMSE), Mean Absolute Error (MAE), Relative Error (RE), and Nash-Sutcliffe Efficiency (NSE) can be used for this comparison (Yang et al., 2014). This comparison is crucial for detecting defects and inconsistencies in datasets (Shem Juma and Kelonye Beru, 2021). Once validated, statistical models can be used to predict future crop yields under various climate change scenarios (Shem Juma and Kelonye Beru, 2021). Overall, crop yield prediction using statistical models is typically associated with uncertainties of both natural and anthropogenic origin, indicating that continuous improvement with a focus on external factors that affect crop yield is necessary (Shem Juma and Kelonye Beru, 2021).

6.2.4 MODELING LIMITATIONS

Although crop modeling is a useful approach for predicting complex processes under different scenarios, uncertainty remains an important topic of debate among researchers (Pasquel et al., 2022). Identifying and managing these uncertainties can improve the performance of crop models and provide more accurate guidance to policymakers for developing adaptation strategies (Chapagain et al., 2022). Sensitivity analysis is the most common method for quantifying the contribution of uncertainty in model inputs to uncertainty in model outputs (for more details on this type of analysis, see Salciccioli et al., 2016). Typically, the uncertainty of a given crop model arises from a combination of errors in the model itself (e.g., equations), model inputs (e.g., soil, climate, and crop data), and model parameters (Pasquel et al., 2022). As such, uncertainties vary with the type of crop model used (GGCMs, PBMs, or statistical models). GGCMs are highly sensitive to soil conditions and crop management; therefore, insufficient global data on soil and crop management can lead to significant uncertainties in model simulation results (Folberth et al., 2019). Additionally, uncertainty in the crop calendar dataset used in GGCMs is a major factor affecting model simulation results (Li et al., 2022). Furthermore, there is an urgent need for better representation of environmental issues such as soil degradation (e.g., soil salinity, erosion, acidification) in GGCMs (Folberth et al., 2019). With respect to uncertainties associated with PBMs, studies (e.g., Chapagain et al., 2022) have shown that uncertainty arises from uncertainty in climate (e.g., rainfall), soil, initial conditions, and crop management data (e.g., sowing date). Statistical models may also have errors and deficiencies (Martínez et al., 2020). For example, they lack the ability to predict crop yields under extreme future climate conditions (e.g., higher temperatures than any historical year, elevated CO_2 concentrations) (Soussana et al., 2010). Additionally, the CO_2 fertilization effect (Dhakhwa et al., 1997) is not well represented in these models (Kogo et al., 2019; Martínez et al., 2020). To improve the performance of statistical models, this effect should be included (Martínez et al., 2020). The CO_2 effect can be estimated in the field using standard methods; for example, Sugasti and Pinzón (2020) provide a good example of this estimation. Overall, Chapagain et al. (2022) called for the development of standard procedures for estimating crop model uncertainty and evaluating simulation results.

6.3 CLIMATE CHANGE EFFECTS ON AGRICULTURAL PRODUCTION BY 2050

Climate change by 2050 is expected to pose a threat to global agricultural production by causing significant changes in crop yields and livestock productivity under various GHG emissions scenarios (e.g., Hristov et al., 2020). Numerous previous studies (e.g., Wiebe et al., 2015; Hasegawa et al., 2022; Godde et al., 2021) have used climate, crop, and economic models to assess the global and regional effects of climate change by 2050 on crop yields and livestock productivity. However, results have varied significantly due to differences in models, scenarios, and input data (Wiebe et al., 2015). The following subsections summarize the main findings of these studies.

6.3.1   Potential Global Crop Yields by 2050

According to some estimates, in the absence of effective adaptation measures and if GHG emissions continue at their current rate, global crop yields could decrease by up to 30% by 2050 (Global Commission on Adaptation, 2019). However, it is important to note that the effects of climate change on crop yields by 2050 will vary by country (FAO, 2018). Projections at a global level under a medium GHG emission scenario, using 2011 as a starting point, show that some regions such as Canada, the United States of America (USA), and Russia will experience higher agricultural production by 2050. In contrast, other regions such as India and West Africa could face critical reductions in crop yields by 2.9% and 2.6%, respectively (Figure 6.3) (FAO, 2018). Overall, crop yields are expected to increase in higher latitude countries such as Canada and Russia due to the potential positive effects of higher temperatures, an extended growing season, and CO_2 fertilization. However, lower latitude countries and arid and semiarid countries (e.g., most African countries) will likely face agricultural production losses due to an expected further decrease in precipitation (Rosenzweig et al., 2014; FAO, 2018).

In African countries, where crop yields are already low and issues such as land degradation, pollution, and deforestation are serious challenges, climate change by

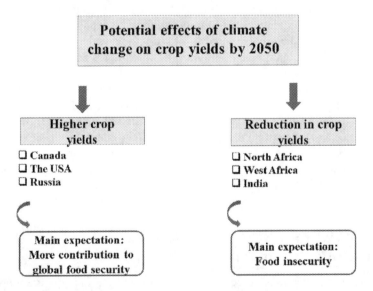

FIGURE 6.3   Potential effects of climate change on crop yields by 2050.

2050 is likely to further reduce agricultural production (FAO, 2018) (Figure 6.3). This could have critical effects on the African population due to their high dependence on agriculture for livelihoods and their limited capacity to adapt to a changing climate (Knox et al., 2012). As a result, food insecurity may intensify in the African continent, particularly in sub-Saharan Africa (Knox et al., 2012). In addition to Africa, Southern Asia could also experience a 10%–16% reduction in crop yields by 2050 relative to the 1990–2015 baseline period and a notable increase in the prices of staple foods under a high GHG emission scenario (Yan and Alvi, 2022). Compared to the average yield over 1981–2010, some major crops such as maize and wheat in European countries may also see decreases of about 22% and 49%, respectively, due to climate change by 2050 (Hristov et al., 2020). However, in Northern European countries (e.g., the Netherlands, Finland), the negative effects of climate change on crop yields may be partially offset by the CO_2 fertilization effect and changing precipitation regimes (Hristov et al., 2020).

Although low and medium GHG emissions by 2050 may lead to an enhancement of crop yields in some countries such as Canada (Carew et al., 2017) and Northern Europe (Hristov et al., 2020), the majority of previous studies (e.g., Nelson et al., 2014; Hristov et al., 2020; Hasegawa et al., 2022) have confirmed that high GHG emissions without adaptation measures are likely to reduce crop yields by 2050. For instance, relative to the 2000 baseline year, Nelson et al. (2014) found that the global average crop yield could decrease by up to 11% by 2050 under a high-emission representative concentration pathway (RCP 8.5). Limited CO_2 fertilization effects in 2050, together with the extreme climate conditions of the RCP 8.5 scenario (e.g., an increase in temperature), could be the main reasons for this potential decrease in global average crop yield by 2050 (Nelson et al., 2014). It is estimated that a 1°C increase in temperature may result in a 10–20% reduction in global crop yield (Elahi et al., 2022). Moreover, a recent study conducted by Wing et al. (2021) found that climate change could reduce global crop yields by up to 12% in 2050. However, it is important to note that the projected decrease in crop yields due to climate change by 2050 is larger for some crops than others (see Table 6.1). At a global level, under a runaway climate change scenario by 2050, it is expected that the yield of the top four global crops (wheat, maize, rice, and soybean) would decrease by −13%, −14%, −16%, and −30% between 2010 and 2050, respectively (Müller and Robertson, 2014). In addition to these four global crops, other crops such as groundnut (Müller and Robertson, 2014), potato (Raymundo et al., 2018), tomato (Cammarano et al., 2022), and banana (Varma and Bebber, 2019) could also face significant yield losses (see Table 6.1).

The projections summarized above reflect the urgent need for quick intervention through the implementation of adequate adaptation measures. Even in regions that are expected to experience a decrease in crop yields by 2050, if adaptation measures are adequately implemented, the negative effects of climate change could be minimized. Hasegawa et al. (2022) revealed that the decrease in global average crop yield could be reduced from a rate of −11% without adaptation to −4.6% with effective adaptation. This confirms that future crop yield estimates that do not include possible adaptation measures may exaggerate the negative effects of climate change on agricultural production.

TABLE 6.1

Projected Changes in Global Productivity for the Top Four Global Crops (Wheat, Maize, Rice, and Soybean) and Others (Groundnut, Potato, Tomato, and Banana) by 2050

Crop Type	Baseline Period	Project Period	Climate Change Scenario	Change in Global Productivity (%)	Reference (Data Source)
Wheat	2010	2050	High GHG emission	−13	Müller and Robertson (2014)
Maize	2010	2050	High GHG emission	−14	Müller and Robertson (2014)
Rice	2010	2050	High GHG emission	−16	Müller and Robertson (2014)
Soybean	2010	2050	High GHG emission	−30	Müller and Robertson (2014)
Groundnut	2010	2050	High GHG emission	−21	Müller and Robertson (2014)
Potato	1979–2009	2055	High GHG emission	−6	Raymundo et al. (2018)
Tomato	1980–2009	2050	High GHG emission	−6	Cammarano et al. (2022)
Banana	1970–2000	2050	Medium GHG emission	−56	Varma and Bebber (2019)

GHG: greenhouse gas.

6.3.2 POTENTIAL GLOBAL LIVESTOCK AND FISHERIES PRODUCTIVITY BY 2050

Livestock products (e.g., meat, milk, and eggs) and fisheries are critical components of global food security (Swanepoel et al., 2010). Specifically, they provide 33% of global protein and 17% of global calories consumed (Cheng et al., 2022). Moreover, the livestock sector provides livelihoods and income for many people worldwide, with about 1.1 billion people employed in the sector (Hurst et al., 2005). Climate change factors such as higher temperatures, increasing precipitation variability, and more frequent extreme climate events by 2050 are expected to affect the productivity and quality of livestock both directly through heat stress affecting animal mortality and productivity and indirectly through impacts on water availability, grassland species distribution, and diseases (Godde et al., 2021) (Figure 6.4). Overall, livestock production will face significant pressure under climate change conditions due to competition for land and water resources and food security at a time when it is most needed (Thornton, 2010; Rojas-Downing et al., 2017). This means that the ability of the livestock sector to support livelihoods and meet the rising demand for productivity could be threatened to varying degrees depending on the region and climate change scenarios (Godde et al., 2021; Brouillet and Sultan, 2023).

Projections of livestock and fisheries productivity by 2050 under climate change scenarios at a global scale are still limited (Rojas-Downing et al., 2017). Among the few available projections, Boone et al. (2018) predicted a reduction in global livestock productivity by 2050 under a high GHG emission scenario (RCP8.5) by an average rate of 7.5%–9.6% due to reduced herbaceous production, with the highest impact projected for sub-Saharan western African countries (about −46%). However, large increases in annual productivity are expected for some higher latitude countries

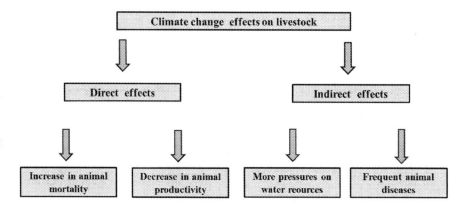

FIGURE 6.4   Direct and indirect effects of climate change on livestock.

(e.g., a 21% increase in productivity in Canada) (Boone et al., 2018). Climate change by 2050 may also have a significant effect on rangelands (i.e., areas for feeding livestock) (Boone et al., 2018). For instance, Godde et al. (2020) showed that 74% of the global rangeland area is projected to experience a reduction in average herbaceous biomass, with the largest regional reduction expected for Oceania and the highest increase expected for Europe due to the fertilization effect of increasing CO_2 concentration.

In addition to its effect on livestock productivity, climate change may also affect global fish production. Barange et al. (2014) and Lam et al. (2016) found that global fish production could decrease by 5%–10% by 2050. These examples confirm that future climate change by 2050 may threaten the livelihoods of millions of people worldwide, particularly in low-income countries where people strongly depend on livestock productivity for food security (Brouillet and Sultan, 2023). Therefore, there is an urgent need to promote adaptation measures that can enhance livestock productivity under climate change.

6.4   FOOD DEMAND AND SECURITY BY 2050

Under the projected climate conditions in 2050 (e.g., increased temperature and rainfall variability), relative to 2010, overall agricultural production must increase by up to 56% (van Dijk et al., 2021) to meet food demands for at least an additional 2.5 billion people (from about 6.9 billion in 2010 to at least 9.4 billion by 2050, United Nations, 2019) (Figure 6.1). The production of major global staple crops (e.g., wheat, barley, maize, soybean, and rice), livestock products, and fisheries will need to increase effectively by 2050 (Table 6.2). More specifically, using the year 2000 as a starting point, global barley production will need to rise by at least 54% to meet projected demand for food, feed, and industrial purposes (Kruse, 2011). Additionally, relative to 2010, global wheat production will need to increase by up to 56% (i.e., an additional production of 359 Mt) to meet increased food demand (Kettlewell et al., 2023). Furthermore, compared to the period from 1961 to 2008, global maize, rice, and soybean production would need to increase by about 67%, 42%, and 55%,

TABLE 6.2

Required Increase in Global Crop Production, Livestock Productivity, and Fisheries by 2050 (Examples)

		Baseline Period	Project Year	Required Increase (%)	References
Major crops	Wheat	2010	2050	+56	Kettlewell et al. (2023)
	Maize	1961–2008	2050	+67	Ray et al. (2013)
	Rice	1961–2008	2050	+42	Ray et al. (2013)
	Soybean	1961–2008	2050	+55	Ray et al. (2013)
	Barley	2000	2050	+54	Kruse (2011)
Livestock products	Global livestock productivity	2000	2050	+100	Garnett (2009)
	Meat production	2006	2050	+76	Alexandratos and Bruinsma (2012)
	Milk production	2006	2050	+63	Alexandratos and Bruinsma (2012)
Fisheries	Global fisheries productivity	2000	2050	+100	Garnett (2009)

respectively, which is far below what is needed to meet projected demands in 2050 (Ray et al., 2013). On the other hand, relative to 2000, global fisheries and livestock productivity will need to improve by 100% by 2050 (Garnett, 2009). More precisely, Alexandratos and Bruinsma (2012) showed that relative to 2006, global meat production is projected to increase by 76% (from 258 Mt in 2006 to 455 Mt in 2050) and milk production by 63% (from 664 Mt in 2006 to 1,077 Mt by 2050).

In light of the findings discussed in the third section of this chapter, it appears that meeting global food demand for about 9.4 to 10.1 billion people by 2050 and ensuring global food security will be a significant challenge for the international community. The challenge is to maintain a balance between sustainable agricultural productivity, food security, and environmental preservation (Wright et al., 2012). Recent food security assessments (Searchinger et al., 2019) have shown that the global agriculture sector is capable of meeting global food requirements by 2050, but this will only be possible if adequate adaptation measures such as improving crop breeding, enhancing soil and water management, reducing GHG emissions from agricultural activities, and continuing to improve crop yield productivity through sustainable methods are successfully implemented in the coming years.

6.5   CONCLUSIONS

This chapter reviewed the future relationship between global agricultural productivity and climate change by 2050. It is clear from this review that under some severe climate change scenarios by 2050 (i.e., high GHG emissions without adequate intervention), climate change could have catastrophic multidimensional effects on global agricultural productivity. While increasing temperatures and CO_2 levels by 2050 may be beneficial for crops in some countries such as Northern European countries, Canada, and Russia, potential changes in global or regional climate patterns due to climate change by 2050 could critically affect crop yields and livestock productivity directly and indirectly in other countries, mainly lower latitude countries, leading to severe food security risks. Globally, there is an urgent need to promote all adaptation measures that can enhance agricultural productivity under climate change in the coming years to avoid global food insecurity by 2050. Comprehensive and coordinated studies on the real-time effects of climate change on global agricultural productivity and corrective policy decisions are highly needed in the coming years for better management of food risks at a global scale.

7 European Agriculture Under Climate Change
Past Observations and Projected Figures by 2050

7.1 INTRODUCTION

The agriculture sector in the European Union (EU) plays a crucial role in ensuring food security both within and outside the EU due to its high productivity in cereal, vegetables, wine, fruits, sugar, and livestock products (EC, 2021). For example, EU countries produce 67% of global cereal (Bindi and Olesen, 2011), and France is a leading producer of sugar beet with yields reaching 13.7 tons sugar ha^{-1} (Heno et al., 2018). Additionally, countries such as France and Germany have significant livestock production (milk, meat, eggs, and seafood), which is a critical component of food security (Ritchie et al., 2017). In 2008, EU countries accounted for approximately one-fifth of global meat production (Knox et al., 2016). Agriculture is practiced on approximately 39% of Europe's land area (EEA), highlighting the importance of this sector and its potential impact on the landscape and natural environment (Knox et al., 2016). Furthermore, agriculture provides employment opportunities for rural populations in the EU (Iglesias and Garrote, 2015).

European agriculture, which provides numerous services, is currently threatened by factors such as population growth and climate change (Zhao et al., 2022). The rapid increase in the EU's population over the past five decades has led to unprecedented demand for food (van Zanten et al., 2016), placing pressure on the agriculture sector to ensure sustainable food production. Climate change has also had significant impacts on agriculture in Europe, with warming trends, droughts, heat waves, flooding, and changes in precipitation patterns (Copernicus Climate Change Service, 2022). These changes have resulted in decreased crop yields and nutritional quality, loss of agricultural land in some regions, changes in soil quality (e.g., due to erosion and salinity), increased production costs, rising food prices, and food insecurity (Dadson et al., 2010; Webber et al., 2020). Without intervention, these effects are projected to continue and even worsen (Carozzi et al., 2022), posing a serious threat to European agriculture and global food security in the coming decades (Carozzi et al., 2022).

Understanding the relationship between climate change and agricultural production is essential for effective management of agricultural areas under changing climate conditions (Fisher et al., 2012). Previous studies using agro-ecosystem models, climate projections, and field observations (e.g., Bindi and Olesen, 2011; Olesen and Bindi, 2002)

DOI: 10.1201/9781003404194-9

have provided valuable insights into the impacts of climate change on agriculture in Europe. However, there are still significant gaps in our understanding of these effects, particularly in the coming decades (Hristov et al., 2020). Investigating past and future trends in European agriculture under climate change is crucial for maintaining productivity, identifying mitigation constraints, and developing policies to support farmers in adapting to global warming (Puertas et al., 2023). In this context, the main objectives of this chapter are to investigate the historical effects of climate change on agriculture in Europe and to analyze projected trends up to 2050.

7.2 OBSERVED CLIMATE CHANGE EFFECTS ON EUROPEAN AGRICULTURE

In recent decades, climate change has had varying impacts on agriculture in Europe, depending on factors such as agroclimatic conditions, land use intensity, infrastructure availability, and political and economic policies (Knox et al., 2016). Overall, the critical effects of climate change on European agriculture include decreased production, loss of suitable areas for traditional crops, and reduced livestock productivity (Figure 7.1). The following subsections provide further details on these impacts:

7.2.1 EFFECTS ON AGRICULTURAL PRODUCTION AND LIVESTOCK PRODUCTIVITY

The impacts of climate change on agricultural production represent both an ecological and food security challenge (Bai et al., 2022). Numerous studies have examined the effects of climate change on agricultural production in Europe (e.g., Moore and Lobell, 2015; Beillouin et al., 2020; Conradt et al., 2023). These studies have consistently found that the observed decrease in European agricultural production over the past three decades is primarily due to the effects of climate change on crop yields, growing season conditions, and water demand.

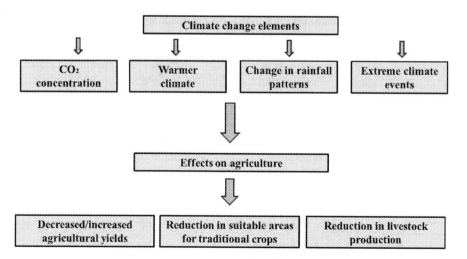

FIGURE 7.1   The observed primary effects of climate change on European agriculture.

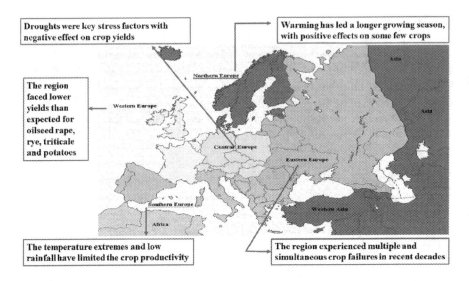

FIGURE 7.2   Regional variations in the observed effects of climate change on European crop yields.

1. Climate change effects on European crop yields:

   In recent decades, warming trends have had significant impacts on crop yields in Europe (Beillouin et al., 2020). As shown in Figure 7.2, these impacts have varied by region (Beillouin et al., 2020). In some areas of Northern Europe, warmer temperatures have had positive effects on certain crops that have benefited from a longer growing season (Olesen, 2016; Biemans, 2018). However, overall, the effects of climate change on crop yields in Europe have been mostly negative, with warmer conditions reducing yields for many crops (Biemans, 2018). For example, production-weighted continent-wide yield decreases of 2.5% and 3.8% were observed for wheat and barley, respectively, during 1989–2009 (Moore and Lobell, 2015; Faye et al., 2023). Extreme climate events such as droughts and heatwaves have also reduced crop yields across much of Europe (Beillouin et al., 2020), with an estimated loss of 7.3% in crop production during 1991–2015 due to these events (Brás et al., 2021). An analysis of yield data from 17 European countries for nine key crops (barley, maize, oats, oilseed rape, potatoes, triticale, rye, sugar beet, and wheat) using machine learning models found that both temperature and precipitation extremes had negative impacts on yields over much of Europe between 2000 and 2018, with considerable regional variation (Beillouin et al., 2020). Overall, the study found that large yield losses were observed in many parts of Northern and Eastern Europe while enhanced yields were observed in Southern Europe due to increased spring rainfall (Beillouin et al., 2020). These findings are consistent with those of Moore and Lobell (2015), who reported that extreme climate events since 1990 have decreased average wheat and barley yields in much of Europe by 2.5%. In Central Europe, the 2018–2019 drought event resulted in unprecedented agricultural production losses of −20 to −40% for staple crops such as maize, rice, wheat, and potatoes compared to 2015–2017 records (Conradt et al., 2023). Schmitt et al. (2022) found that one drought day can decrease winter wheat yield by up to 0.76%. Negative impacts of climate change on crop yields have also been observed in Western Europe, with lower than expected yields for oilseed rape, rye, triticale, and potatoes in many areas (Beillouin et al., 2020).

2. Climate change and change in growing season:
   Temperature is a key abiotic factor affecting the phenological development of crops (Ceglar et al., 2019), and gradual warming in many European regions over the past few decades has extended the growing season for several crop varieties (Jeong et al., 2011; Ceglar et al., 2019). Chmielewski and Rötzer (2001) found that a 1°C increase in annual average temperature can extend the growing season by approximately 5 days. On average, the growing season for European crops has increased by about 10 days since 1992 (EEA, 2019), with more pronounced increases in Northern and Eastern Europe than in Western and Southern Europe (Jeong et al., 2011). This prolonged growing season has brought benefits to Northern Europe, including the introduction of new crop species, more extensive use of double-crop rotation, and improved timber growth in forests (Zhao et al., 2022). However, adaptation measures such as plant breeding and pest control are necessary to address challenges such as the spread of weeds, insect pests, and diseases (Zhao et al., 2022). In addition to extending the growing season, warmer temperatures in most of Europe have also contributed to significant changes in growing season conditions, such as the beginning and end of the season. A study by Aalto et al. (2022) found that the average growing season in Northern Europe began 15 days earlier during 1990–2019 compared to previous years. These changes in the timing and length of the growing season have led to significant alterations in plant life cycles (Vogel, 2022). For example, in Southern Europe, there has been a trend toward earlier leaf unfolding, flowering, and fruiting for deciduous shrubs and crops (Vogel, 2022). European farmers have been adapting to these changes by adjusting their cultivation timing, agronomic management practices, and crop species and cultivar selection (Olesen et al., 2011).

3. Climate change effects on crop water demand:
   Climate change in Europe has implications not only for crop development and growing season conditions but also for crop water demand (Wriedt et al., 2009; Soares et al., 2023). However, there are limited observations of the effects of climate change on crop water demand in Europe. Available studies (e.g., Wriedt et al., 2009; Soares et al., 2023; van der Velde et al., 2010; Zajac et al., 2022) have shown that climate change has led to increased crop water demand in large parts of Southern and Eastern Europe over the past three decades, while a decreasing trend has been observed in parts of North-Western Europe. Irrigation is mainly practiced in Mediterranean countries, where up to 80% of abstracted water is used for irrigation annually (Eurostat, 2021). However, due to higher temperatures, lower rainfall, and increased frequency of droughts and heatwaves, irrigation is becoming more popular in traditionally nonirrigated areas of Central and Northern Europe where water resources are not yet a constraint for agricultural production (Greaves and Wang, 2016; Zajac et al., 2022).

In 2015, livestock production systems occupied 28% of the EU's land surface, corresponding to approximately 65% of agricultural land (Leip et al., 2015), highlighting the importance of these systems in Europe. Compared to studies on crop production, there are fewer studies on the effects of climate change on livestock productivity at the European level (e.g., Özkan et al., 2016; Dono et al., 2016; Pasqui and Giuseppe, 2019). Like crop yields, the livestock sector in Europe has also experienced productivity losses due to climate change (Morignat et al., 2014). Meat, milk, and egg production have decreased significantly over the past few decades due to higher temperatures, increased precipitation variability, and more frequent extreme climate events such as droughts and flooding in many European countries (Pasqui and Giuseppe, 2019). The increased frequency of extreme events, particularly heatwaves and droughts, in many European regions has negatively affected animal health through changes in eating behavior, disease distribution, and poor mental health, leading to increased mortality rates and reduced productivity (Dono et al., 2016; Pasqui and Giuseppe, 2019). For example, heatwaves in France between 2003 and 2006 increased cattle mortality by 12% to 24% (Morignat et al., 2014).

7.2.2 EFFECTS ON AGRICULTURAL AREAS

In recent decades, many European countries have experienced a decline in total agricultural area due to factors such as modernization, land use change, intensification of agriculture, and climate change-induced land degradation processes such as erosion and salinity (Hatna and Bakker, 2011). Climate change can accelerate soil erosion (Panagos et al., 2018) and soil salinity (Daliakopoulos et al., 2016), both of which contribute to the decline in agricultural area in Europe. These processes pose significant threats to European agriculture by reducing soil fertility and causing losses in production, damage to plantations, and reduction of available planting area (Daliakopoulos et al., 2016; Panagos et al., 2018). On an EU scale, it is estimated that the 12 million ha of agricultural land facing severe erosion lose approximately 0.43% of their crop productivity annually, resulting in an economic loss of €1.25 billion per year (Panagos et al., 2018). Soil salinity, which restricts agricultural production, affects approximately 3.8 to 4 million ha in Europe (Daliakopoulos et al., 2016). The highest levels of soil salinity are mainly observed along the coastlines of Mediterranean countries such as Italy, Spain, and France due to mechanisms that facilitate the accumulation of salts in the soil, including higher temperatures, rapid evaporation, excessive irrigation, rising sea levels, flooding, and saline intrusion of groundwater (De Waegemaeker, 2019). Soil salinity has also been detected in some areas of Northern Europe (Gould et al., 2021), although studies on the effects of salinity and climate change on agriculture in this region are limited (Gould et al., 2021). The decline in agricultural area coupled with decreased crop yields could result in further reductions in crop production (EC, 2021), highlighting the need for adaptation measures to prevent further development of erosion and salinity processes in European agricultural areas.

7.3 PROJECTED CLIMATE CHANGE EFFECTS ON EUROPEAN AGRICULTURE BY 2050

7.3.1 Projected Crop Yields and Livestock Productivity

Reliable projections of crop yields under climate change are essential for developing policy plans to address food security (Agnolucci and De Lipsis, 2020). Several studies in Europe (e.g., Olesen et al., 2011; Sutton et al., 2013; Hristov et al., 2020) have investigated the potential trends in farm crop yields by 2050. These studies have projected both negative and positive trends, with significant regional differences due to variations in climatic factors such as temperature and precipitation (Hristov et al., 2020). Figure 7.3 summarizes the projected effects of climate change on crop yields by 2050 for different sub-regions of Europe (Northern, Central, Southern, Eastern, Western). A positive trend (i.e., potential increases in crop yields) is expected only for Northern Europe while a negative trend (i.e., decreases in crop yields) is projected for the other subregions (Olesen et al., 2011).

- The potential positive effect of climate change on crop yields in Northern Europe around 2050 is attributed to several factors, including an expected increase in precipitation, a shorter growing season, and warmer climatic conditions (Hristov et al., 2020). By 2050, most key crops (e.g., wheat, soybean) are projected to have higher yields (Hristov et al., 2020). For example, relative to the average yield over 1981–2010, wheat yield in the region is expected to increase by 5% to 16% around 2050 (Hristov et al., 2020). Additionally, warmer climatic conditions are projected to increase soybean productivity from an average value of 1.2 t ha⁻¹ over 1981–2010 to 1.6 t ha⁻¹

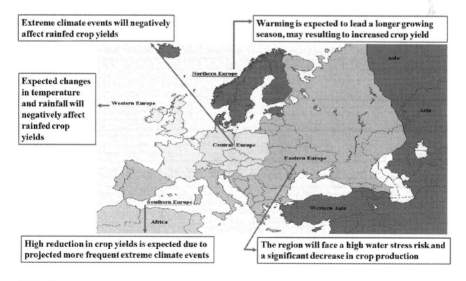

FIGURE 7.3 Projected climate change effects on European crop yields by 2050 (under without adaptation scenario).

in 2050 under a low greenhouse gas (GHG) emission scenario (i.e., RCP4.5 scenario), even without considering the effects of elevated CO_2 concentration (Guilpart et al., 2022). Furthermore, Northern Europe's agriculture is expected to benefit from a longer and warmer growing season in the coming years. New crop varieties may be introduced in areas previously limited by low temperatures or shorter growing seasons (Olesen et al., 2011).

• The largest decrease in crop yields is projected to occur in Southern Europe (Hristov et al., 2020). Under all GHG emission scenarios (from low to high emissions), significant decreases in yields are projected for key crops such as maize, wheat, potato, soybean, and sugar beet (Olesen et al., 2011). For example, relative to the average yield over 1981–2010, the average maize yield under a high GHG emission scenario (RCP8.5) is projected to decrease by 22% by 2050, while wheat yield is expected to decrease by up to 49% (Hristov et al., 2020). This is not surprising given that the region is expected to experience higher temperatures and lower rainfall by 2050 (Copernicus Climate Change Service, 2022). Extreme climate events (e.g., droughts) can significantly decrease crop yields in the coming decades, especially when they occur during the most sensitive stages of crop growth (e.g., Webber et al. 2018). For instance, in Southern Portugal, more frequent high-temperature events are projected to reduce wheat yield (relative to 1981–2010) by 14% over 2021–2050 under both low (RCP4.5) and high (RCP8.5) GHG emissions (Yang et al., 2019). Additionally, some crops that currently grow mostly in Southern Europe may become more suitable further north or at higher altitudes in the south (Olesen et al., 2011). Crop yield losses in the region could be reduced in the coming decades through tailored adaptation measures such as changing crop varieties, improving irrigation practices, and enhancing soil management practices (Hristov et al., 2020).

• Due to a projected increase in average temperature (by 1.5 to 2°C) and significant changes in precipitation patterns by 2050, agricultural areas in Eastern Europe are expected to face high water stress risk and a significant decrease in crop production if no adaptation measures are adopted (Sutton et al., 2013). For example, under a medium climate change scenario, maize yield is projected to decrease by 10% in Moldova and by up to 54% in Macedonia (relative to current yield) around 2050 (Sutton et al., 2013). Vegetable yields are also expected to decrease significantly by 2050 (Sutton et al., 2013).

• In Central and Western Europe, where agricultural productivity is already high (Olesen et al., 2011), projected changes in temperature and precipitation are expected to negatively affect the yields of rainfed crops such as wheat, barley, and maize (Biemans, 2018).

Projected climate change effects on crop yield also show large differences between crops. For example, in Southern Europe, wheat yield is projected to decrease by up to 49% by 2050 (relative to the average yield over 1981–2010) while in Northern Europe, some of the negative climate change effects on wheat yield may be partially offset by higher levels of atmospheric CO_2 concentrations and changing precipitation regimes (Hristov et al., 2020). Crops with statistically significant yield variations include wheat, maize, potato, and sugar beet (Knox et al., 2016). These results suggest that the development of climate adaptation strategies in the agriculture sector should consider location and crop-specific sensitivity to climate change. Livestock production in Europe is also projected to decrease in the coming decades, particularly in Southern European countries due to an expected increase in the frequency of extreme events such as heatwaves and droughts (EEA, 2019).

7.3.2 PROJECTED CHANGE IN GROWING SEASON CONDITIONS

Over the past three decades, the growing season for several European crops has extended. This trend is projected to accelerate throughout most of Europe in the coming decades, particularly in Northern and Eastern Europe. This will facilitate the northward expansion of warm-season crops into agricultural areas that are currently unsuitable (Ceglar et al., 2019; Ruosteenoja et al., 2020). Under a high greenhouse gas emission scenario (RCP8.5), the length of the growing season in Europe is projected to increase by 1.5 to 2 months by 2100 relative to the reference period of 1971–2000 (Ruosteenoja et al., 2016). However, rising temperatures and more frequent extreme climate events (e.g., droughts) may reduce the suitability of some currently productive agricultural areas in Southern Europe for certain crops, shifting production to Northern Europe (e.g., King et al., 2018; Ceglar et al., 2019). In Northern Europe, projected increases in annual-average temperature and changes in precipitation patterns by 2050 are expected to have positive effects on agriculture, allowing farmers to introduce new crop species, increase crop yields, and potentially have multiple harvests from the same plot (e.g., Peltonen-Sainio and Jauhiainen, 2020). However, it is important to note that Ceglar et al. (2019) reported that the extension of the growing season in Northern and Eastern Europe could also have some disadvantages, such as increased risk of late frost and early spring and summer heatwaves.

7.3.3 PROJECTED LAND DEGRADATION

Recent studies (e.g., Panagos et al., 2021; Gould et al., 2021) have indicated that the risk of salinity and erosion processes in Europe will increase dramatically due to more frequent extreme climate events, such as intense rainfall and droughts. This is expected to result in further land degradation and a reduction in the total agricultural area. According to Panagos et al. (2021), the total agricultural area in the EU and the United Kingdom is projected to decrease by approximately 7 million hectares by 2050. Panagos et al. (2021) also projected the effects of climate change and land use change on future rates of soil erosion by water within the agricultural areas of the EU under low (RCP2.6), medium (RCP4.5), and high (RCP8.5) greenhouse gas emissions scenarios. They estimated that by 2050, soil loss due to water erosion could increase by up to 13% relative to the 2016 baseline. In some European regions, projected future climatic conditions, such as flooding and sea level rise, are also expected to pose a greater threat of salinity to agricultural systems (Gould et al., 2021). These results underscore the need for stronger soil conservation measures against erosion and salinity to avoid the worst-case scenario by 2050.

7.4   CONCLUSIONS AND RECOMMENDATIONS

Over the past three decades, climate change has had a significant impact on agriculture in Europe, resulting in lower productivity, changes in growing season conditions, livestock production losses, and reductions in suitable areas for traditional crops. This issue poses a threat to agriculture in the medium term (by 2050) in the majority of European regions, with critical consequences for food security. As such, it is essential to promote adaptation strategies that can enhance European agricultural productivity under changing climatic conditions. Adapting to climate change must be made a top priority for the EU's agriculture sector. While there is currently a great deal of knowledge about the effects of climate change on European agriculture, there are still some gaps that need to be addressed. For example, further investigation into the effects of rainfall on growing season conditions in water-limited areas is highly recommended. It is also important to consider surrounding natural ecosystems when investigating the effects of climate change on agricultural systems at the local scale. Additionally, conducting more in-depth analyses at higher resolution for key European agricultural areas is recommended. Many large-scale studies on the effects of climate change on crop yields at the European scale have made simplifying assumptions (e.g., about crop management), which may introduce uncertainties into estimated effects. Therefore, efforts should be made to reduce these uncertainties as much as possible to ensure a more accurate assessment of the effects of climate change on European agriculture.

8 Climate Change Effects on African Agriculture
Observation and Projected Trends by 2050

8.1 INTRODUCTION

Climate change, manifested through warming, droughts, floods, and changes in precipitation patterns, can have significant effects on agriculture (Bhattacharyya et al., 2020). It can profoundly impact various soil properties, such as soil carbon, moisture, and salinity, as well as freshwater availability, plant/crop yield, and livestock production (Bhattacharyya et al., 2020). For example, significant decreases in soil organic carbon (SOC) stocks often occur as a result of climate warming (Crowther et al., 2016) and inappropriate land management practices (Keel et al., 2019). Additionally, warming trends, changes in precipitation patterns, prolonged droughts, and decreased water availability may all contribute to reduced agricultural productivity (Arora, 2019). Moreover, increased temperature and rainfall variability could lead to the emergence of numerous livestock diseases, potentially resulting in a substantial reduction in livestock production (Rojas-Downing et al., 2017).

Given the immense value of agriculture worldwide, there are national, regional, and international concerns regarding the effects of climate change on agriculture (Emediegwu et al., 2022). Despite the vulnerability of African agriculture to climate change (Emediegwu et al., 2022), these concerns have not received sufficient attention. Previous studies (e.g., Pereira, 2017; Emediegwu et al., 2022) have shown that agriculture in the majority of African countries is highly susceptible to climate change due to its rainfed nature, its economic significance as a key activity on the continent, the financial constraints of smallholder farmers, poverty, and low adoption of adaptive technologies. Furthermore, climate change is expected to have adverse effects on African agriculture in the near future (Girvetz et al., 2019). In many parts of the continent, climate change is projected to decrease agricultural production, particularly affecting the productivity of staple crops such as rice, wheat, and maize. This could have significant impacts on the livelihoods and food security of many farmers, affect livestock production, and lead to further impacts on key soil properties (e.g., Li and Fang, 2016; Pereira, 2017; Bhattacharyya et al., 2020). These potential effects will exacerbate the vulnerability of Africa's agriculture to climate change. Consequently, African agriculture must become more adaptive to climate change (Girvetz et al., 2019). Predicting the effects of climate change on African agriculture under future climate scenarios is essential for establishing effective adaptation measures (Sharma and Sharma, 2022). This is crucial in supporting policymakers in taking decisive steps toward climate change adaptation across the entire African continent. In this context, the primary objectives of this chapter are (1) to examine observed historical effects of climate change on African agriculture and (2) to review projected trends of these effects by 2050.

8.2   OBSERVED HISTORICAL CLIMATE CHANGE EFFECTS ON AFRICAN AGRICULTURE

Over the past three decades, all African countries have experienced rapid increases in temperatures, changes in precipitation patterns, and alterations in extreme climate events, such as more frequent droughts. These changes have resulted in reduced agricultural productivity (Pereira, 2017), losses in livestock production (e.g., Tabler et al., 2021), and severe soil degradation (e.g., Chapman et al., 2021). The following subsections provide further details on these effects of climate change:

8.2.1   Effects on Agricultural Productivity

The effects of climate change on crop production are typically assessed using process-based biophysical modeling, which involves the sequential use of General Climate Models (GCMs) and dynamic crop growth simulation models such as DSSAT, CropSyst, AquaCrop, and CROPWAT (Zinyengere et al., 2013). Using this methodology, previous studies (e.g., Exenberger et al., 2014; Ogundari and Onyeaghala, 2021) have shown that among climate change factors such as precipitation, snow, temperature, and wind, rainfall has the greatest impact on African agricultural productivity. Specifically, rainfall consistently increases agricultural productivity in Africa. For example, a study conducted by Ward et al. (2011) revealed that cereal growth across Sub-Saharan Africa (SSA) increases significantly with increasing rainfall. Additionally, Adhikari et al. (2015) showed that a one-inch decrease in rainfall amount could reduce maize yield by 9% (23%). Researchers have differing estimates regarding the effects of temperature on agricultural productivity. For instance, Exenberger et al. (2014) reported an insignificant negative impact of temperature on agricultural production in SSA. However, Ward et al. (2011) demonstrated that temperature can have a significant impact on agricultural productivity, with cereal yield across SSA decreasing with increasing temperatures.

In Africa, where agricultural production is heavily dependent on rainfall, it is essential to understand how changes in rainfall patterns affect it (Ogundari and Onyeaghala, 2021). Accordingly, numerous studies have been conducted to analyze rainfall patterns in Africa and their effects on agriculture. For example, Chapman et al. (2020) found that reduced rainfall rendered large parts of SSA unsuitable for multiple staple crops. Additionally, Shikwambana et al. (2021) analyzed rainfall patterns between 1960 and 2018 in Limpopo Province, South Africa, and found that increased recurrence of drought and rainfall variability during that period resulted in water scarcity, decreased crop production, and recurrent food crises. Furthermore, in Nigeria, Haider (2019) showed that changes in rainfall patterns have led to numerous drought and flooding

TABLE 8.1

Climate Change Effects on African Agricultural Productivity

Region	Observed Effects	Key Reference
Western Africa	Climate change has caused an average maize yield reduction of −8%	Roudier et al. (2014)
Northern Africa	Over 1961–2009, a 1% increase in winter temperature resulted in a 1.12% drop in agricultural productivity	Alboghdady and El-Hendawy (2016)
Eastern Africa	Climate change in the region can result in a decrease of 1.2% to 4.5% in agricultural production	Abraha-Kahsay and Hansen (2016)
Central Africa	Climate change and its associated water stress have resulted in a notable decent yield of the key crops	Molua and Lambi (2007)
Southern Africa	In 2015–2016, climate change increased food production losses sharply, leading to 26 million people in the region needing humanitarian assistance	Funk et al. (2018)

events that adversely affect crop production throughout the country. Since crops are highly sensitive to increasing temperatures and changing rainfall patterns (Ayanlade et al., 2021), climate change in Africa has significantly impacted the productivity of commonly grown crops such as wheat, maize, and sorghum. Overall, according to a recent report by the UN's Intergovernmental Panel on Climate Change (IPCC), climate change has decreased agricultural productivity on the continent by more than a third since 1961. Although Knox et al. (2012) identified an average crop yield reduction of −8% across the entire African continent, it is worth noting that the negative effects of recent climate change on crop yield vary significantly by region (Table 8.1):

- According to Nelson et al. (2009), the negative impact of climate change on crop yield is more pronounced in SSA compared to other regions of the continent. Mulungu and Ng'ombe (2020) observed that, at the beginning of the 2000s, maize, rice, and wheat yields in SSA had declined by 10%, 15%, and 34%, respectively, due to climate change.
- Western Africa is considered a major hot spot in terms of climate change and food security (Sultan and Gaetani, 2016). In this region, maize is the crop most affected by climate change, with Roudier et al. (2014) identifying an average maize yield reduction of 8%. This reduction can be attributed to the rapid climate change in the region, characterized by rising temperatures and more frequent extreme climate events, leading to losses in crop yields (Sultan and Gaetani, 2016). Furthermore, Stuch et al. (2021) demonstrated that under high heat stress, maize yields in West Africa do not benefit from increased soil moisture resulting from increased rainfall. Additionally, Sultan et al. (2019) found that altered climate conditions from 2000 to 2009 resulted in regional average yield reductions of 10–20% for millet and 5–15% for sorghum.
- In Northern Africa, heatwaves and droughts over recent decades have increased water stress, reduced the land areas suitable for agriculture,

shortened the length of growing seasons, forced large areas of marginal agriculture out of cultivation, and decreased yields of many crops (Gornall et al.,
2010; Schilling et al., 2020). Heatwaves during crop development phases can
result in fewer organs, decreased light interception due to shortened crop
life, and altered carbon-assimilation processes such as transpiration, photosynthesis, and respiration, leading to substantial decreases in crop yield
(Stone, 2001). Alboghdady and El-Hendawy (2016) investigated the effects
of climate change on agricultural output in North Africa from 1961 to 2009
and found that a 1% increase in winter temperature resulted in a 1.12% drop
in agricultural productivity.

- Eastern Africa has experienced increasing temperatures, shifting rainfall patterns, and more frequent extreme events (Waithaka et al., 2013;
Adhikari et al., 2015; Gebremeskel et al., 2019; Haile et al., 2020).
Collectively, these aspects of climate change have resulted in a significant
decrease in agricultural production in recent years. Abraha-Kahsay and
Hansen (2016) found that climate change in the region could result in a
decrease of 1.2–4.5% in agricultural production. Their study also showed
that mean season temperature and precipitation, as well as growing-
season precipitation variability, have notable effects on crop production.
Specifically, they found a substantial negative effect of growing-season
precipitation variability.

- Climate change has been identified as the primary contributor to severe
water shortages in Central African countries, including Chad, Cameroon,
Central Africa Republic, Gabon, and Equatorial Guinea (ATPS, 2013). This
is particularly evident in the Lake Chad Basin, one of Africa's largest water
bodies, which has experienced a 90% reduction in its water body within the
past 30 years (ATPS, 2013). As key crops such as maize, wheat, and rice
require significant amounts of water for production, the water stress associated with climate change has resulted in decreased yields of these crops in
many Central African countries. This has led to food insecurity and neces-
sitated food aid from organizations such as the World Food Programme
(WFP). For example, in 2005, the WFP organized an emergency operation
to provide 9,500 tons of food to Cameroon to compensate for food shortages (Molua and Lambi, 2007). Given the high dependence on crop yields
for both subsistence and economic purposes in Central African countries,
there is an urgent need to implement adaptation measures to enhance crop
yields under changing climate conditions.

- In Southern Africa, agriculture is predominantly practiced under rainfed conditions (Twomlow et al., 2008), rendering crop production highly
susceptible to the impacts of climate change (Ziervogel et al., 2008). As a
result, numerous studies have been conducted to assess the effects of climate change on crop production in the region (Zinyengere et al., 2013).
For example, a study by Funk et al. (2018) found that climate change has
increased drought-induced food production losses in Southern Africa, leading to a need for humanitarian assistance for 26 million people in the region
during 2015–2016.

TABLE 8.2

Climate Change Effects on Livestock Production (Examples from Africa)

Climate Change Factor	Observed Effects	Example(s)
Increasing temperatures	Decrease in animal production	Temperatures higher than 30°C can decrease poultry production (Ensminger et al., 1990).
	Decrease in animal milk	Temperatures higher than 25°C can increase cow milk production loss sharply.
Shifts in precipitation patterns	Increased frequency and severity of diseases	Rainfall variability can create conditions favorable for survival and transmission of free-living pathogens (Magiri et al., 2020).
Extreme climate events (e.g., droughts)	Increased mortality	Drought in the Serengeti ecosystem (Northern Tanzania) caused a total loss of 70% of livestock (Leweri et al., 2021).

8.2.2 EFFECTS ON LIVESTOCK PRODUCTION

Livestock production plays a crucial role in enhancing food security, improving live-lihoods, generating employment opportunities for rural households, and contributing to economic growth for hundreds of millions of people in Africa (Meltzer, 1995). On average, livestock production accounts for 26% of the agricultural Gross Domestic Product (GDP) across the African continent (Omollo et al., 2020). However, livestock production systems in many African countries have been adversely affected by increasing temperatures, changes in precipitation patterns, frequent extreme climate events, and water shortages (Omollo et al., 2020). These impacts have resulted in decreased feed intake, reduced milk production, and increased mortality rates among livestock (Table 8.2).

- Effects of increasing temperatures
  In many African countries, the lack of knowledge, skills, and technical information on managing livestock under warming conditions has resulted in significant impacts of increasing temperatures on animal growth and productivity (Tabler et al., 2021). For example, temperatures exceeding 30°C can significantly decrease poultry production by reducing feed intake, weight gain, carcass weight, and egg production (Ensminger et al., 1990; Tankson et al., 2001; Tabler et al., 2021). Additionally, higher temperatures increase the risk of disease and mortality among poultry (Tabler et al., 2021). Similarly, warmer climates have been shown to decrease milk production from cattle-based dairy systems in many African regions (Rahimi et al., 2022). An earlier study investigated the relationship between temperature and milk production in semi-arid areas of the Free State province from 1950 to 1999 and found that temperatures above 25°C sharply increased milk production losses. This is due to the fact that increasing temperatures cause cows to drink more and eat less, leading to a significant decrease in both the quantity and quality of cow milk (Summer et al., 2018).

- Effects of changes in precipitation patterns
  Changes in precipitation patterns have been shown to affect livestock production through the transmission of infectious diseases (Magiri et al., 2020; Tabler et al., 2021). Fluctuations between periods of low and high rainfall can create conditions that are conducive to the survival and transmission of free-living pathogens, leading to the emergence of new animal diseases (Magiri et al., 2020; Tabler et al., 2021).
- Effects of extreme climate events
  In many African countries, extreme climate events such as droughts and floods can have significant impacts on livestock production. Floods can create conditions that are conducive to the survival and transmission of free-living pathogens, leading to the emergence of new diseases and increased mortality risk among animals (Magiri et al., 2020). Droughts, on the other hand, can force animals to travel long distances in search of food and water (Kimaro et al., 2018) and can also affect livestock health by causing metabolic disruptions, oxidative stress, and immune suppression (Magiri et al., 2020). This makes animals more susceptible to infectious diseases and increases their mortality risk (Haseeb et al., 2019). For example, in Tanzania, where livestock production is heavily dependent on rainfall, a study by Leweri et al. (2021) found that recurrent drought events over the past decade have resulted in massive losses of livestock, particularly cattle. Similarly, the Borana people of Southern Ethiopia experienced a loss of approximately 42% of all cattle due to severe drought events from 1991 to 1993 (Desta and Coppock, 2002). Also, in the Somali region, drought events during 1990–2000, and 2001–2002/2003 have increased animal death rates by 60% and 80% of the entire cattle population, respectively (Bogale and Erena, 2022). In the areas surrounding the Serengeti Ecosystem in Northern Tanzania, the drought of 2017 was responsible for a total loss of approximately 70% of livestock compared to the livestock count in 2016 (Leweri et al., 2021). Additionally, Archer et al. (2021) reported that droughts in 2020/2021 in the Eastern Karoo region of South Africa had unprecedented impacts on livestock production, including loss of condition among animals, negative implications for reproduction, decreased milk production, and significant loss of income.

8.2.3 EFFECTS ON SOIL RESOURCES

Soil degradation resulting from climate change is a major issue across Africa, with negative impacts on human populations, food productivity, water security, and biodiversity (Borrelli et al., 2017; Chapman et al., 2021). Climate change exacerbates soil degradation through changes in climate variables such as temperature and rainfall (AGNES, 2020). Despite the severity of this issue, there is a lack of up-to-date data on soil degradation caused by climate change at the continental level (AGNES, 2020). According to AGNES (2020), soil degradation affects approximately 46% of Africa's land area and impacts around 485 million people. The continent is currently facing several forms of soil degradation, including erosion and salinity, which diminish soil functions and ecosystem services (Karmakar et al., 2016).

- Soil erosion:
  Despite estimates indicating that Africa has the highest soil erosion rates globally due to high rainfall erosivity and conversion of forest to cropland (Borrelli et al., 2017; Chapman et al., 2021), there are few studies available on historical soil erosion at the continental level (Chapman et al., 2021). Estimating soil erosion in a given agricultural area requires data on climate factors (e.g., rainfall, temperature), soil characteristics (e.g., depth, texture), topographic factors (e.g., slope, landform), and land use and crop management practices (e.g., tillage practices, crop rotation) (Salumbo, 2020) (Figure 8.1). By integrating these data into software packages such as the Revised Universal Soil Loss Equation (RUSLE), Soil and Water Assessment Tool (SWAT), and Soil Loss Estimation Model for Southern Africa (SLEMSA), researchers can effectively estimate soil erosion and develop appropriate soil conservation measures (Salumbo, 2020) (Figure 8.1). Using these models, soil erosion has been estimated in some African regions. For example, Wynants et al. (2019) found that Eastern Africa is the most eroded region in Africa, resulting in a significant decrease in agricultural productivity (Fenta et al., 2019). This critical soil erosion issue is primarily caused by steep topography, fragile soils, and intense rainfall (Wynants et al., 2019; Chapman et al., 2021). In Ethiopia, approximately 1 billion ton of topsoil is lost annually due to soil erosion, costing the country approximately 3% of its agricultural GDP (Doukkali et al., 2018). Climate change has been identified as a major contributor to this significant soil loss (AGNES, 2020).

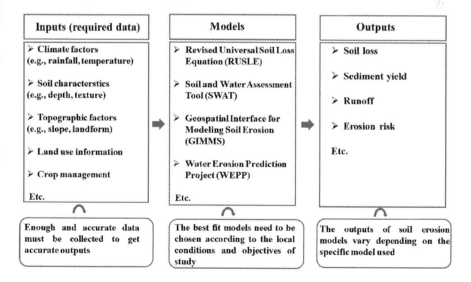

FIGURE 8.1   Data and models required to estimate soil erosion.

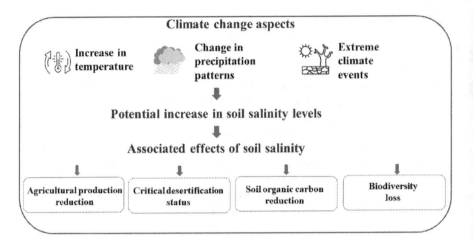

FIGURE 8.2 The effects of climate change and soil salinity on agriculture, desertification, biodiversity, and soil organic carbon.

- Soil salinity:

  Soil salinity has emerged as a critical environmental issue in Africa due to its negative impacts on agriculture, biodiversity, desertification, and soil biogeochemistry (Haj-Amor et al., 2022) (Figure 8.2). Climate change has been identified as a major contributor to soil salinity (Corwin, 2020). At the continental level, the process of soil salinization is most pronounced in North Africa, where more than 150,000 hectares of land in Morocco, Libya, and Egypt have become saline (AGNES, 2020). Low rainfall and high temperatures are the primary factors increasing soil salinity in this region (Sakadevan and Nguyen, 2010). In addition to other factors such as irrigation with saline water, inefficient drainage, and upward movement of salts from shallow groundwater, climate change-related phenomena such as sea level rise and associated saltwater intrusion, higher temperatures, more intense rainfall events, and frequent droughts and heatwaves can also exacerbate soil salinity in agricultural areas (Corwin, 2020). During drought periods, upward water movement resulting from capillary rise from shallow groundwater can cause salts to accumulate in the soil. Over time, this progressive accumulation of salts can significantly increase soil salinity (Corwin, 2020). For example, in Morocco (North Africa), where approximately 0.5 million hectares of agricultural land are affected by salinity (Oumara and El Youssfi, 2022), the recurrence of droughts from 1984–2018 has been identified as a major factor contributing to soil salinity in the country (Rafik et al., 2022). In Southern Tunisia (North Africa), where date palm (*Phoenix dactylifera*) is a key horticultural crop, droughts over the past decade have forced farmers to use highly saline water to irrigate their date palms. This has increased soil salinity to critical levels (> 6 dS m^{-1}) and led to the abandonment of date palm plantations in many agricultural areas (Haj-Amor et al., 2016). In coastal agricultural areas, rising sea levels have increased the frequency, duration, and spatial extent of seawater intrusion into fresh groundwater aquifers (Weissman and Tully, 2020), increasing the risk of soil salinity in these regions (Thiam et al., 2021). For example, in Senegal, out of the 3.8 million hectares of agricultural land in the country, 1.7 million hectares are affected by salinity at the national level due to seawater intrusion (Thiam et al., 2021).

8.3 PROJECTED CLIMATE CHANGE EFFECTS ON AFRICAN AGRICULTURE

According to recent forecasts by the United Nations (UN), the population of Africa is projected to increase from approximately 1.3 billion in 2020 to 2.4 billion in 2050. This increase is expected to exacerbate the pressure on African agriculture to meet growing food demands, which are already impacted by climate change (Emediegwu et al., 2022). Achieving food security for 2.4 billion people in Africa by 2050 will depend heavily on the ability of agriculture to feed the rapidly growing population and mitigate the adverse effects of climate change (Sultan et al., 2019). As such, it is imperative to understand the future effects of climate change on African agricultural productivity in order to prevent a food crisis and prepare for unexpected impacts through effective adaptation measures. This section summarizes the projected effects of climate change on African agriculture.

- Overall, without timely intervention in the coming years, yields of key crops for African populations, such as wheat, maize, sorghum, and millet, are projected to decrease in the coming decades due to potential warming trends (Carleton, 2022). A warming of 2°C could lead to a decrease in crop yields across SSA by 10% while warming beyond 2°C could result in a decrease of 20% (Carleton, 2022). The situation could worsen if the warming trend exceeds 3°C, rendering all present-day cropping areas for maize, millet, and sorghum in Africa unsuitable for cultivation (Carleton, 2022). This would have devastating consequences for Africa's food security as sorghum and millet are both vital staple cereals in most African diets (Carleton, 2022).
- Under a high GHG emissions scenario, yields of most crops in SSA, including wheat, maize, and rice, are projected to decrease by 2050 (Adhikari et al., 2015; Serdeczny et al., 2017; Chapman et al., 2021). As such, there is an urgent need for improved crop management practices in response to climate change, such as transitioning to more heat and drought-resistant crops.
- In North Africa, an expected decrease in annual rainfall by 2050, coupled with a potential rise in temperatures and an increase in the number of extreme hot days, will result in increased demand for irrigation water and severe water shortages (Haj-Amor et al., 2020). For example, in Southern Tunisia, where temperatures are projected to increase sharply, the irrigation water requirement for date palms is expected to increase from an average of 1,459 mm per year in 2018 to 1,950 mm per year by 2050 (Haj-Amor et al., 2020). As the region is already facing growing drought events, farmers may struggle to find enough water to keep their date palms alive.

88 Sustainable Agriculture

- In Eastern Africa, rising temperatures in the coming decades are expected to have varying impacts on crop yields. In high-elevation areas, such as mountainous regions in Ethiopia and Kenya, crop yields may increase. However, in lowland areas, higher temperatures are likely to result in decreased crop yields and increased risk of water stress (Adhikari et al., 2015).
- In Western Africa, without appropriate adaptation measures, climate change is projected to result in a mean reduction in crop yields of 11% by 2050 (Roudier et al., 2011; Sultan et al., 2019). As such, agricultural investments in both conventional high-yielding technology and adaptation measures are recommended to ensure food security in the face of aggressive climate change (Iizumi et al., 2017; Sultan et al., 2019).
- Climate change is also projected to have negative impacts on livestock production systems in the coming decades (Pereira, 2017). For example, regions such as North Africa and Southern Africa, which are expected to become drier, may become more vulnerable to livestock losses due to prolonged drought periods (Niang et al., 2014; Pereira, 2017). Additionally, climate change is expected to reduce the availability of feed for livestock. In many East African countries, for instance, the availability of maize stover per head of cattle is projected to decrease by 2050 (Thornton et al., 2010; Pereira, 2017). Furthermore, climate change in Africa by 2050 is likely to increase pressures from pests, weeds, and diseases, with detrimental effects on both crops and livestock (Niang et al., 2014; Pereira, 2017).

8.4 CONCLUSIONS

Agriculture in Africa is particularly vulnerable to climate change due to its reliance on rainfed systems, its significant economic role, and its relatively low adaptive capacity to climate impacts. Studies reviewed in this chapter confirm that various aspects of climate change, such as rising temperatures, frequent droughts, and changes in rainfall patterns, are increasingly impacting agricultural productivity across the African continent. Without timely intervention, these impacts are projected to worsen by 2050. Additionally, African agriculture is exposed to a range of stressors, such as soil degradation (e.g., erosion, salinity), increasing demand for food, and rapid population growth, which interact with climate change in complex ways. As such, agricultural adaptation measures in Africa should be developed based on social, physical, and environmental approaches.

9 Past and Future (2050) Effects of Climate Change on Asian Agriculture

9.1 INTRODUCTION

The agriculture sector plays a crucial role in driving economic and social development in Asia due to its fundamental contributions to food security, employment, and poverty reduction (Liu et al., 2020). As a result of the green revolution, Asian countries increased their share of global agricultural production from approximately 23% in 1961 to over 44% in 2011 (Alston and Pardey, 2014). In 2010, more than 43% of the total population relied on this sector for their livelihoods. Additionally, the sector has helped to reduce hunger in developing Asian countries from approximately 550 million in 2001 to about 389 million in 2019 (FAOSTAT, 2022). On average, agriculture in Asia contributes about 66% of global agricultural Gross Domestic Product (Mendelsohn, 2014).

Climate change poses a significant threat to Asian agriculture (Pickson et al., 2023). Over the past few decades, various climate change-related phenomena, including droughts, heatwaves, erratic and intense rainfall patterns, and floods, have had adverse effects on this sector (Mendelsohn, 2014). These phenomena have critically impacted crop yields, food security, soil resources, and water availability. For example, water resources in Central Asia have experienced significant shortages over the past three decades due to climate change-induced temperature rise and low rainfall (Yang et al., 2019). The Food and Agriculture Organization (FAO) estimated that climate change-related disasters resulted in an economic loss of approximately 47 billion USD in the Asian agriculture sectors between 2003 and 2013. Additionally, many Asian countries have observed reductions in yields of major crops such as rice, wheat, and maize due to changes in temperature and rainfall patterns (Aryal et al., 2020). Climate change has also contributed to significant changes in growing season conditions such as length and start day. For instance, a study by Abbas et al. (2017) found that between 1980 and 2014, the length of spring maize growing seasons in Pakistan decreased by approximately 4.6 days per decade while the length of autumn maize growing seasons increased by 3 days per decade, critically affecting crop yield (Aryal et al., 2020).

Previous studies have projected that the effects of climate change on Asian agriculture will worsen in the absence of adequate intervention in the coming years (Knox et al., 2012; Li et al., 2022). For example, Knox et al. (2012) projected that climate change will reduce the yield of major crops such as rice, wheat, and maize in

South Asia by approximately 8% by 2050. Additionally, relative to the year 2000, the price of these major crops in 2050 is estimated to increase by 2.5 times (Nelson et al., 2009). Under a moderate GHG emissions scenario, it is expected that the Asian Water Tower, which covers an area of about 3×10^6 km^2, comprises abundant solid and liquid freshwater reservoirs, and provides water to nearly 2 billion people (Yao et al., 2022), may lose a substantial portion of its water storage by 2050, resulting in critical effects on water availability and food security. While several studies have explored the impact of climate change on Asian agriculture (Auffhammer et al., 2012; Holst et al. 2013; Wang et al. 2014; Aryal et al., 2020), the majority have focused on specific countries such as China and India and particular crops such as rice, corn, and soybean. Comprehensive analyses across all Asian regions remain limited (Mendelsohn, 2014; Pickson et al., 2023), despite the entire Asian region being more susceptible to climate change risks than any other region (McKinsey Global Institute, 2020; Pickson et al., 2023). In this context, the main objectives of this chapter are to review past effects of climate change on agriculture across all Asian regions and to analyze trends in these effects until 2050.

9.2   OBSERVED CLIMATE CHANGE EFFECTS ON ASIAN AGRICULTURE

In Asia, the direct impact of climate change on agriculture over the past few decades has manifested as a decline in crop yields and a reduction in water resources (Du et al., 2021) (Figure 9.1). The following subsections provide further details on these effects of climate change.

9.2.1   EFFECTS ON CROP YIELDS

Climate plays a crucial role in determining crop yields; therefore, changes in climate such as droughts, high temperatures, and intense rainfall can affect crop productivity

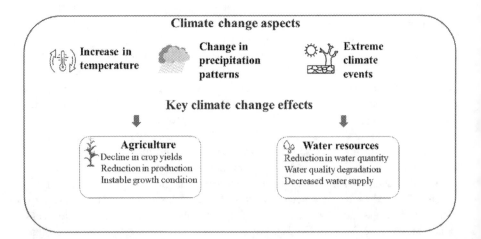

FIGURE 9.1   Key effects of climate change on Asian agriculture and water resources.

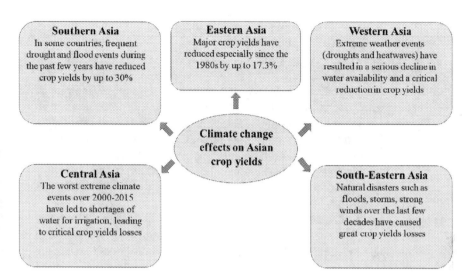

FIGURE 9.2   Observed climate change effects on Asian crop yields over the past few decades.

in agricultural areas (Abdul Rahman, 2018). Investigating crop yield responses to historical climate change is essential for coping with anticipated changes in temperature and precipitation (Bhatt et al., 2014). In Asia, historical effects of climate change on crop yields over the past few decades have shown both positive and negative trends (Aryal et al., 2020). Increases in crop yields, that is, positive trends, as a result of climate change have only been identified in a few Asian regions (Bhatt et al., 2014). For instance, Bhatt et al. (2014) found that an increase in average temperature by 0.42°C per decade during 1968–2008 in high-elevation areas of Nepal led to an enhancement of rice yield at a rate of 17.7 kg ha^{-1} year^{-1}. However, a wide range of yield losses has been observed in almost all subregions of Asia, including Eastern, South-Eastern, Central, Southern, and Western Asia (Ray et al., 2019; You et al., 2022; Santini et al., 2022; Pickson et al., 2023) (Figure 9.2).

- In Eastern Asia, including countries such as Japan, South Korea, and China, the majority of available studies have shown that changing climatic conditions in the region, such as flooding, severe heatwaves, and droughts, have decreased yields of major crops over the past few decades, particularly since the 1980s (Liu et al., 2012; Lee et al., 2022). For example, in South Korea, a water-deficient country, episodes of droughts between 2013 and 2017 decreased yields of major crops such as rice and wheat by 17.3% (Lee et al., 2022; Park et al., 2022). Similarly, in China, changes in average climate variables during 1961–2010 resulted in a decrease in rice yield by up to 12% (Liu et al., 2012).
- In South-Eastern Asia, including countries such as Brunei, Indonesia, Malaysia, Thailand, Vietnam, and the Philippines, climate variability and extreme climate events such as floods and droughts over the past few decades have resulted in loss of crop harvest and decreased productivity (Abdul Rahman, 2018;

Huynh et al., 2020). For example, floods that occurred in Southern Malaysia in 2006 reduced the production of crude palm oil by more than 26% that year (Abdul Rahman, 2018). In Vietnam, natural disasters such as floods, storms, strong winds, and heatwaves between 2011 and 2018 caused significant crop yield losses, particularly for rice, coffee, pepper, avocado, and cashew (Huynh et al., 2020). Overall, it is estimated that these natural disasters resulted in a reduction in crop productivity of approximately 20%–30% (Huynh et al., 2020). Due to the frequent recurrence of natural disasters, Vietnam is ranked by the United Nations Development Organization (UNDO) as one of the countries most vulnerable to climate change (Huynh et al., 2020). Between 1970 and 1990, climate variability in the Philippines resulted in rice production losses of more than 80% (Lansigan et al., 2000; Suresh et al., 2021).

- Agriculture in the Central Asia region, including countries such as Kazakhstan, Kyrgyzstan, Tajikistan, Turkmenistan, and Uzbekistan, is highly sensitive to climate change due to its strong dependence on water resources (Umirbekov et al., 2022). Between 2000 and 2015, the region experienced some of the worst extreme climate events, primarily droughts and heatwaves (Patrick, 2017), which led to shortages of water for irrigation and critical crop yield losses (Aitekeyeva et al., 2020). For instance, a drought that occurred in Kazakhstan in 2012, particularly during the critical period of wheat growth, reduced wheat yield by up to 36% that year while a drought in Kyrgyzstan in 2008 decreased it by 22% relative to usual yields (Aitekeyeva et al., 2020). Therefore, Central Asian countries should promote the cultivation of more drought-resistant crop varieties on a larger scale and improve irrigation water management under dry conditions (Narbayep and Pavlova, 2022).

- In Southern Asia, including countries such as Bangladesh, India, Pakistan, Nepal, and Sri Lanka, it has been observed that climate change phenomena such as rainfall variability, warming trends, droughts, and floods have critically decreased yields of all major food crops (Aryal et al., 2020). For example, in India, even with adaptation measures in place, rainfall variability in the country led to a decrease in wheat yield by approximately 5% between 1981 and 2009 (Gupta et al., 2017). In the Northwestern and Southwestern regions of Bangladesh, frequent recurrence of drought and flood events in recent years has reduced average crop production by 25%–30% (Hossain et al., 2022). Overall, increasing maximum and minimum temperatures between 1972 and 2009 in Bangladesh decreased rice yields by approximately 37% (Sarker et al., 2012). Between 1974 and 2007, Sri Lanka experienced drastic changes in rainfall patterns, droughts, and temperature, resulting in a reduction in crop area of more than 800,000 ha during that period (DMC, 2009; Suresh et al., 2021).

- In Western Asia, including countries such as Armenia, Azerbaijan, Bahrain, Georgia, Iraq, Jordan, Kuwait, and Saudi Arabia, agricultural production is highly vulnerable to increasing temperatures and decreased rainfall, particularly in agricultural areas that require irrigation (Ortas and Lal, 2013). In recent decades, extreme weather events such as droughts and heatwaves have become more frequent and prolonged in the region, decreasing water availability and critically reducing crop yields (Ortas and Lal, 2013). In the majority of Western Asia countries, including Saudi Arabia and Iraq, rainfall variability and frequent extreme climate events between 1990 and 2010 led to a critical decrease in crop yields (Ortas and Lal, 2013).

Based on the comprehensive description provided above, it is evident that climate change over the past few decades has contributed to massive crop yield losses across all subregions of Asia. Climate-driven extremes such as droughts, heatwaves, intense rainfall, and floods are the primary reasons behind these losses. Crop yield losses, particularly in cropping systems that play a crucial role in food security, have created food security issues and challenges in the region (Cai et al., 2016; Habib-ur-Rahman et al., 2022). Therefore, there is an urgent need to promote resilient agricultural practices to ensure sustainable agricultural productivity and food security under climate change (Habib-ur-Rahman et al., 2022).

9.2.2 Effects on Water Resources

According to Alexander and West (2011), the high dependence on water for agriculture in Asia exacerbates the critical impact of climate change on water availability. From 1985 to 2000, total water withdrawals in the region increased by approximately 25% (329,160 GL) to meet the substantial water demands of agriculture. In countries such as Uzbekistan, Turkmenistan, and Tajikistan, the situation is particularly dire as annual water withdrawals exceed available renewable sources. Climate change factors including rising temperatures, rainfall variability, and extreme weather events (e.g., droughts and heatwaves) have further reduced water availability in many Asian countries (Balasubramanian and Saravanakumar, 2022). For example, droughts from 1997 to 2006 significantly reduced water availability in the southern Murray–Darling Basin, Australia's largest river system (Alexander and West, 2011). Additionally, a warming trend since 1980 (~0.44°C per decade) and precipitation variability have altered the balance of the Asian Water Tower, one of Asia's largest water resources, resulting in increased water scarcity in some regions (Yao et al., 2022). In India, the world's largest user of groundwater with an annual withdrawal of approximately 230 km^3, rising maximum temperatures over recent decades have critically decreased groundwater availability in some regions (Balasubramanian and Saravanakumar, 2022).

9.3 FUTURE CLIMATE CHANGE EFFECTS ON ASIAN AGRICULTURE

According to previous studies (e.g., Thomas et al., 2013; Tao et al., 2022; Umirbekov et al., 2022), the primary impacts of future climate change on Asian agriculture will include (1) significant changes in agricultural productivity due to increased temperature and rainfall variability; (2) reduced crop yields in many parts of Asia by the 2050s; (3) difficulty in arranging planting seasons and crop structure due to extreme and irregular climate phenomena such as early, late, or unseasonal rains, resulting in substantial damage to agricultural production; (4) increased fluctuation in crop production and exacerbation of food insecurity in some parts of Asia; and (5) reduced water availability in many parts of Asia.

TABLE 9.1

Examples of Future Change in Crop Yields by 2050 across Asia (Relative to the Last Years)

Crop	Country/Subregion	Change in Yield by 2050 +: Increase Trend –: Decrease Trend	Reference
Wheat (*Triticum aestivum* L.)	China/Eastern Asia	–6.3%	Tao et al. (2022)
	Sri Lanka/Southern Asia	–6.5%	Aryal et al. (2020)
	Bangladesh/Southern Asia	–32%	World Bank (2020)
	Armenia/western Asia	–8%	USAID (2017)
Rice (*Oryza sativa* L.)	Japan/Eastern Asia	–17.2%	Iizumi et al. (2011)
	Cambodia/South-Eastern Asia	–2.3%	Thomas et al. (2013)
	Thailand/South-Eastern Asia	+28%	Teng et al. (2016)
	Bangladesh/Southern Asia	+17%	World Bank (2020)
	India/Southern Asia	–6% to –8%	Aryal et al. (2020)
Maize (*Zea mays* L.)	Uzbekistan/Central Asia	–22%	Thomas et al. (2021)
	Pakistan/Southern Asia	–10.7%	Yan and Alvi (2022)
	India/Southern Asia	–30%	Aryal et al. (2020)
Barley (*Hordeum vulgare*)	South Korea/Eastern Asia	+20%	Choi et al. (2021)
	Uzbekistan/Central Asia	+0.7%	Thomas et al. (2021)

9.3.1 POTENTIAL CROP YIELDS BY 2050

Several studies have utilized climate, crop, and economic models to estimate the potential effects of 2050-climate on crop yields (e.g., Thomas et al., 2013; Aryal et al., 2020; Tao et al., 2022). These studies confirm significant changes in crop yields for Asia by 2050 relative to current levels (Table 9.1).

Sivakumar and Stefanski (2011) reported that crop yields could increase by up to 20% in Eastern and South-Eastern Asia, while decreasing by up to 30% in Central and Southern Asia by 2050, even when accounting for the direct positive physiological effects of CO_2 (Figure 9.3).

- In Eastern Asia, projected changes in average climate by 2050 are expected to significantly alter crop yields in many countries (Sivakumar and Stefanski, 2011). Due to uncertainties in crop yield projections, climate change scenarios, CO_2 fertilization effects, and adaptation options, previous studies have reported both positive and negative impacts of climate change on crop yields across the region (e.g., Japan, South Korea, China). For example, in China, Tao et al. (2022) projected a 6.3% reduction in wheat yield by 2050 under a medium (RCP4.5) GHG emission scenario relative to the baseline period (1986–2005), without accounting for the CO_2 effect. However, when considering the CO_2 effect, wheat yield was estimated to increase by an average of 5.7% by 2050 (Tao et al., 2022). In South Korea, Choi et al. (2021) found that a CO_2 concentration of up to 400 ppm is favorable for

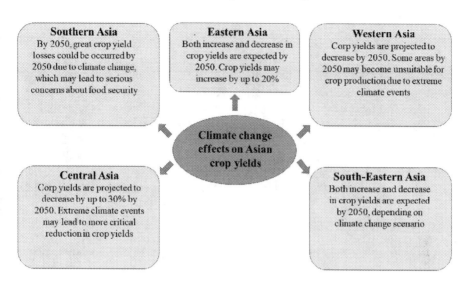

FIGURE 9.3 Projected climate change effects on Asian crop yields by 2050.

increasing soybean (*Glycine max*) and barley (*Hordeum vulgare*) yields. Specifically, compared to the baseline period (1999–2011), soybean and barley yields were projected to increase by approximately 4% and 20%, respectively, by 2050 under the RCP4.5 scenario. In northern Japan, Iizumi et al. (2011) projected a 17.2% increase in rice (*Oryza sativa* L.) yield by 2050 relative to the average yield in the 1990s.

- South-Eastern Asia (e.g., Cambodia, Indonesia, Malaysia, Thailand, Viet Nam, Philippines) accounts for approximately 26% and 40% of global rice production and exports, respectively (Yuan et al., 2022). With global rice demand projected to increase by 30% by 2050 (Yuan et al., 2022), it is crucial to understand the potential impacts of climate change on rice production in this region. Depending on the climate change scenario and location, rice yields may either increase or decrease by 2050 (Thomas et al., 2013). For instance, in Cambodia, one of the countries most vulnerable to climate change, Thomas et al. (2013) projected a 2.3% decrease in rice yields by 2050 relative to the 1950–2000 baseline period. In contrast, northeastern Thailand may experience an increase in rice yields of up to 28% by 2050 relative to current levels, depending on GHG emissions and their effects on rainfall and evapotranspiration (Teng et al., 2016).

- In Central Asian countries (Kazakhstan, Kyrgyzstan, Tajikistan, Turkmenistan, and Uzbekistan), urgent action is required to mitigate the potentially severe agricultural damages from climate change. Thomas et al. (2021) investigated the effects of climate change on yields of major crops in Central Asia under a high (RCP8.5) GHG emission scenario and found that the region is likely to experience greater climate shocks by 2050. Specifically, compared to the year 2000, maize yield in Uzbekistan is projected to decrease by up to 22%, while potato yield in Kazakhstan may decrease by less than 18% by 2050. Spring barley yield in Uzbekistan is projected to increase by 0.7%, while in Tajikistan it may decrease by 2.8%. The study also found that winter wheat yield in Tajikistan is expected to increase by 2050, while the decline projected for Uzbekistan is modest.

- In Southern Asia, the annual average maximum temperature is projected to increase by up to 2.6°C by 2050, and heat waves are expected to become more common, with heat-stressed areas in the region potentially increasing by up to 21% (Ortiz et al., 2008; Tesfaye et al., 2017; Aryal et al., 2020). As a result of heat stress, approximately half of the Indo-Gangetic Plains, the primary food basket of South Asia, may become unsuitable for wheat production by 2050. These long-term changes in climate conditions could reduce wheat production in the region by up to 50% by 2050 (de Fraiture et al., 2007; Ali et al., 2017), raising serious concerns about food security. Overall, Knox et al. (2012) estimated that 2050-climate in Southern Asia would reduce the average yield of major crops (e.g., wheat, rice, and maize) by 8%. However, regional differences in crop yield responses to projected climate change have been identified. For example, the average wheat yield loss in Sri Lanka is projected to be only 6.5% by 2050 (Aryal et al., 2020), while in Bangladesh it may reach approximately 32% due to irregular rainfall patterns (World Bank, 2020; Chowdhury et al., 2022). Rice yield in Bangladesh is projected to decrease by up to 17% by 2050 (World Bank, 2020), while in India it may decrease by only 6%–8% (Aryal et al., 2020). In Pakistan, maize yield is projected to decrease by approximately 10.7% by 2050 (Yan and Alvi, 2022), while in India it may decrease by up to 30% (Aryal et al., 2020). It is important to note that extreme climate events in South Asia over the coming decades are likely to have significant impacts on crop production, with the most severe effects expected in mountainous regions (Sengar and Sengar, 2014).
- In Western Asian countries (e.g., Armenia, Azerbaijan, Bahrain, Iraq, Jordan, Kuwait, and Saudi Arabia), agricultural production is heavily reliant on irrigation (Ortas and Lal, 2013). By 2050, rising temperatures and extreme weather events such as droughts and heatwaves are expected to increase crop water requirements and result in significant declines in crop yields such as wheat, alfalfa, apricot, grape, and potato compared to current levels (USAID, 2017). For example, in Armenia, rising temperatures and water shortages by 2050 may reduce wheat yields by approximately 8% relative to current levels (USAID, 2017).

9.3.2 WATER AVAILABILITY BY 2050

Projections of the effects of climate change on water availability can inform the improved management of water resources for agricultural production (Du et al., 2021). Globally, climate change is projected to reduce water availability by 2050, while water use for agriculture is estimated to increase by approximately 19% (UN-Water, 2013). In arid, semiarid, and subhumid regions of Asia, the situation in 2050 could be significantly worse if no adaptive measures are taken due to the combined effects of climate change, population growth, and changing socio-economic conditions. Even a small increase in temperature (e.g., 1.5°C) can severely affect water resource availability in Asia (Vinke et al., 2017). Depending on projected changes in climate conditions by 2050, water availability may increase slightly in Eastern Asia (by approximately 5%), while other subregions of Asia may experience significant declines (e.g., Du et al., 2021).

- Central Asia, with its strong dependence on water resources, is highly sensitive to climate change and may experience critical water shortages. Water resources in this region are primarily provided by transboundary rivers, which play a crucial role in various agricultural activities (Manning et al., 2018). Umirbekov et al. (2022) projected water availability in Central Asia's major rivers (e.g., Murgap, Tedzhen, Surkhandarya, Kashkadarya, Zarafshan, Talas, Syrdarya, and Chu) for the period 2040–2069 relative to the 1975–2005 baseline period and found that under a high GHG emission scenario (RCP 8.5), these rivers may experience a decline in water volumes due to a potential reduction in average annual precipitation. By 2050, water volumes may decrease by 5% in the Syrdarya river and by as much as 15% in the Amudarya river. These projections highlight the urgent need for Central Asian countries to collaborate on ensuring sustainable management of the region's water resources by 2050.
- In Western Asia, the majority of countries (e.g., Armenia, Azerbaijan, Bahrain, Iraq, Jordan, Kuwait, and Saudi Arabia) experience arid and semi-arid climates and have limited water resources. Climate change is projected to exacerbate water shortages in this region by placing additional stress on available water resources (UNEP, 2017). Only four out of twelve countries in this region exceed the water scarcity limit of 1,000 m^3 per person per year (UNEP, 2017). Climate change is projected to critically impact water availability and quality in Western Asian countries (UNEP, 2017). For example, in Iraq, a projected decrease in rainfall by 3% and an increase in average temperature by more than 3°C by 2050 will reduce freshwater availability by 59%, resulting in severe water scarcity levels (Sofer, 2017). In Armenia, higher temperatures and decreased annual rainfall are projected to decrease the water volume of Lake Sevan (the country's largest freshwater lake) by more than 50 M m^3 in 2030 relative to current levels (USAID, 2017). In Saudi Arabia, Chowdhury and Al-Zahrani (2013) estimated that an increase in temperature by up to 4.1°C by 2050 may increase agricultural water demands by 15% to maintain current levels of agricultural production and may result in water quality deterioration.
- In Southern Asia, the Ganges-Brahmaputra-Meghna river system, the Indus River Basin, and the Helmand River Basin are the largest surface water resources in the region. These resources, along with groundwater, play a crucial role in irrigation and agricultural production (Lacombe et al., 2019). However, these rivers are prone to flooding due to intense monsoonal rains, resulting in widespread damage to agricultural areas through erosion and landslides (Davis and Hirji, 2019). Over the coming decades, higher

temperatures and more erratic rainfall patterns are expected to increase agricultural water demands, placing additional pressure on these major rivers. This pressure is expected to occur even in countries projected to experience increases in precipitation (e.g., Bangladesh, Bhutan, India, and Sri Lanka) unless significant improvements are made in water-use efficiency (Davis and Hirji, 2019). Without adequate intervention, the situation may worsen by 2050 as the region may experience a decrease in groundwater recharge by about 10% due to climate change (Clifton et al., 2010).

9.4   CONCLUSIONS

Climate change is a significant issue for agriculture in Asia. Changes in climate conditions have already critically impacted agriculture across all subregions of Asia (Eastern, South-Eastern, Central, Southern, and Western Asia), resulting in decreased agricultural productivity and serious impacts on water resources. Potential future changes in climate by 2050 may further affect crop production, water resources, and agricultural development in Asian countries. Therefore, a comprehensive assessment of climate change effects and adaptation options is essential for these countries. Integrating comprehensive risk assessments and risk reduction measures into decision-making processes could help develop dynamic adaptive policy pathways to cope with climate change and its associated effects on the agriculture sector.

10 Agricultural Production in North America under Climate Change by 2050

Case Studies from the United States and Canada

10.1 INTRODUCTION

North American countries, particularly the United States and Canada, are among the world's strongest agricultural producers, with significant production of maize, wheat, barley, soybeans, fruits, vegetables, and various livestock products (Wang et al., 2015; Bonti-Ankomah et al., 2017). The United States is a world leader in agricultural production and the second-largest exporter of agricultural commodities, reflecting its significant contribution to global food security (Wang et al., 2015). Major exported crops include wheat, soybeans, tree nuts, animal feeds, livestock products, and other agricultural products (USDA, 2022). In 2019, the agriculture sector contributed approximately $1.1 trillion to the USA's Gross Domestic Product (GDP) (USDA, 2022) and provided employment for nearly 22 million people in various roles such as on-farm jobs, agricultural product transport, food production, and food service (USDA, 2022). In Canada, agriculture is practiced over an estimated area of 64 M ha in 2016, with prime farmland concentrated mainly in the western Prairie Provinces, Ontario and Quebec (Qian et al., 2019). The agriculture sector contributed 4% to Canada's GDP in 2016, worth approximately $111 billion (Bonti-Ankomah et al., 2017), and played a key role in meeting domestic and global food demand (Sarkar et al. 2018) while providing employment for approximately 12% of the total workforce in 2016 (Bonti-Ankomah et al., 2017). Due to its significant agricultural potential, Canada can play a crucial role in achieving the global Zero Hunger Sustainable Development Goal (SDG) mandate by 2030 (Zerriffi et al., 2023). Additionally, Canada is a significant contributor to global food exports. According to Agriculture and Agri-Food Canada, in 2010, it was the world's fifth-largest agricultural food exporting country after the European Union (EU), the United States, Brazil, and China (AAFC, 2011).

Agricultural production is highly dependent on climate variables such as temperature, rainfall, wind, and humidity. Unfavorable climate conditions, such as failed rains and episodic drought events, can result in serious land damage and significant declines in crop yields (Sivakumar and Hansen, 2007). Over the past few decades, agricultural production in both the United States and Canada has faced numerous challenges due to climate change and climate-related natural disasters (Motha and Baier, 2005; Paraschivu and Olaru, 2020). Climate change factors such as increasing temperatures, rainfall variability, droughts, and floods can negatively impact crop production through abiotic stresses such as heat and salinity issues (Paraschivu and Olaru, 2020) and biotic stresses such as increased insect and weed pest pressures (Shahzad et al., 2021). Since 1981, the effects of climate change on agriculture in both countries have been observed through substantial changes in the productivity of major crops such as maize (*Zea mays* L.), rice (*Oryza sativa* L.), wheat (*Triticum aestivum* L.), and soybeans (*Glycine max*), as well as significant changes in growing season conditions such as planting and harvesting dates (Morgounov et al., 2018). For example, in the southeastern USA, Sharma et al. (2022) found that rising maximum temperatures from 1980 to 2020 significantly reduced maize and rice yields by −34% and −8.3%, respectively, while shifts in rainfall over that period significantly decreased wheat yield. This can be attributed to the fact that increasing temperatures can lead to water stress, decreased photosynthetic activity, and disrupted enzyme activities, all of which can negatively impact crop yields (Ben-Asher et al., 2008; Sharma et al., 2022). In Manitoba, Canada, it has been observed that extreme climate events, particularly intense rainfall and drought during 1996–2013, resulted in wheat yield losses of approximately 42% (Carew et al., 2017).

Climate change can adversely affect crop yields (Malhi et al., 2021), making it crucial to assess how major crop yields may change under future climate conditions across the United States and Canada. Understanding these future changes is essential to cope with anticipated changes in temperature and precipitation (Malhi et al., 2021). Numerous studies have been conducted on the future effects of climate change factors such as warming trends, floods, droughts, and rainfall variability on crop yields (e.g., Carew et al., 2017; Qian et al., 2019; Sharma et al., 2022). However, the majority of these studies have focused on a single region, with relatively few providing a comprehensive review of all major crops across the entire United States and Canada. In this context, the overall objective of this chapter is to review recent literature on the effects of climate change on crop yields across the United States and Canada by 2050. The specific objective is to provide a regional-scale overview of all relevant future effects, rather than focusing on a single region.

10.2   CROP YIELDS BY 2050 ACROSS THE UNITED STATES

By 2050, annual average warming over the United States is projected to exceed the potential global average warming. Climate projections for the 48 states estimate that average annual temperatures by 2050 will be 0.5–2.8°C higher than average annual temperatures from 1986–2015 (USGCRP, 2018; EDF, 2022). This expected warming trend, along with changes in precipitation patterns and extreme climate events, such as droughts, heatwaves, and storms, is expected to affect yields of key crops in the country to varying degrees depending on the crop and location (Blanc and Reilly, 2015). Differences in location can result in variations in climatic variables, particularly temperature and precipitation, and their rate of change, leading to significant changes in crop yields (Ginbo, 2022). Quantifying future climate change effects on crop yields allows for the identification of appropriate adaptation measures such as selecting alternative

TABLE 10.1

Examples of the Most Common Crop Models

Model	Full Name	Useful Link (Model Description and Details about Model Developers)
DSSAT	Decision Support Systems for Agrotechnology Transfer	https://dssat.net/
APSIM	Agricultural Production Systems Simulator	https://www.apsim.info/
EPIC	Environmental Policy Integrated Climate model	https://ndcpartnership.org/toolbox/ environmental-policy-integrated-climate-model-epic
CERES	Crop Environment Resource Synthesis	https://geomodeling.njnu.edu.cn/modelItem/3f1e5810-3611-475e-b231-f20edf7715b4
FASSET	Farm Assessment Tool	https://www.fasset.dk/
MONICA	Model for Nitrogen and Carbon in Agro-ecosystems	https://github.com/zalf-rpm/monica/wiki
STICS	Multidisciplinary Simulator for Standard Crops	https://www6.paca.inrae.fr/stics_eng/About-us/Stics-model-overview
WOFOST	World Food Studies	https://www.wur.nl/en/research-results/research-institutes/environmental-research/facilities-tools/software-models-and-databases/wofost.htm

crops to mitigate climate change effects (Chen et al., 2021). Using a range of statistical and crop growth models such as the Decision Support Systems for Agrotechnology Transfer (DSSAT), the Agricultural Production Systems Simulator (APSIM), and the Environmental Policy Integrated Climate (EPIC) model (Table 10.1), several studies (e.g., Hsiang et al., 2017; Obembe et al., 2021; Yu et al., 2021) have quantified the potential effect of climate change on crop yields across the United States by 2050.

These studies have yielded different results but all indicate that the overall effect of climate change on crop yields in the country is projected to be negative. Some crop yield projections by 2050 across the United States are as follows (Figure 10.1):

- According to Hsiang et al. (2017), a projected increase in temperature across the contiguous USA by 2050 is expected to result in a reduction in yields of major crops, including maize (*Zea mays* L.), wheat (*Triticum aestivum* L.), soybean (*Glycine max.*), and cotton (*Gossypium*). For every 1°C increase in temperature and associated changes in precipitation and atmospheric CO_2, yields are expected to decrease by approximately 9%. Yu et al. (2021) revealed that, relative to the 2013–2017 baseline period, climate change by 2050 in the United States is expected to decrease average maize and soybean yields by 39–68% and 86–92%, respectively, depending on the climate change scenario and global climate model used. The largest reduction in maize and soybean yields in the United States would occur without CO_2 fertilization, adequate adaptation measures, and genetic crop enhancement (Zhao et al., 2017).
- According to a report by the Environmental Defense Fund (EDF) (2022), under a medium GHG emission scenario (RCP4.5), climate change by 2050

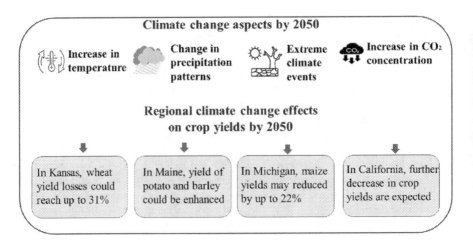

FIGURE 10.1   Effects of climate change on crop yields in the United States by 2050.

is expected to result in a significant decline in freeze days by up to 23% in Kansas, Midwestern USA. This may decrease freeze damage to winter wheat and potentially enhance yields. Additionally, an expected increase in precipitation by approximately 5% under the same scenario may also enhance winter wheat yield. However, it is important to note that an expected increase in the number of hot days (i.e., days with temperatures exceeding 27.8°C) by approximately 96% may critically damage winter wheat yield. As a result of these potential climate factors, winter wheat yields are expected to vary significantly across Kansas by 2050. The eastern part of Kansas may experience an enhancement in winter wheat yields by approximately 5% while other parts of the state may face negative effects on yields due to climate change. Obembe et al. (2021) showed that many areas in Kansas should expect winter wheat yield losses of approximately 16% and 31% under medium (RCP4.5) and high (RCP8.5) GHG emission scenarios, respectively. Freezing temperatures in the autumn and extreme heat in the spring were identified as the main drivers of these expected yield losses. Extreme heat can decrease crop yield by shortening phenological phases (Eyshi Rezaei et al., 2017) and decreasing available soil moisture through increased evapotranspiration (Tooley et al., 2021).

- Tooley et al. (2021) revealed that an increase in CO_2 concentration by 2050 may enhance crop growth and yield of potato (*Solanum tuberosum* L.) and barley (*Hordeum vulgare* L.) in Maine, Northeastern USA. However, it is important to note that the expected yield enhancement due to increased CO_2 may be constrained by higher temperatures and decreased precipitation.
- Liu and Basso (2020) found that the occurrence of drought events during the late phase of the growing season could critically reduce maize and wheat yields by up to 22% and 13%, respectively, under potential climate conditions of a medium GHG emission scenario (RCP4.5) in Southwestern Michigan, United States. In agreement with these results, Wang et al. (2017) also revealed

that exacerbated drought stress could reduce maize (*Zea mays* L.) yield in the St. Joseph River Watershed, located in Northeast Indiana, Northwest Ohio, and South-central Michigan, by 6.4% by 2050 (compared to the baseline period of 1991–2010).

- Heavy rainfall and flood events can have negative effects on crop yield in many regions of the United States. Intense rainfall can delay planting, result in physical damage to crops, degrade agricultural fields, increase soil moisture, and reduce crop yields (Tooley et al., 2021). For instance, in the northern Great Plains (Montana, Nebraska, North Dakota, South Dakota, and Wyoming), winter and spring rainfall by 2050 is expected to increase relative to the 1971–2000 average, with a notable increase in the number of days with heavy rainfall (USGCRP, 2014). These potential climate conditions may lead to many agricultural areas in the northern Great Plains becoming too wet, which may increase erosion intensity, nutrient runoff, delay planting, and affect yields of dominant crops such as alfalfa (*Medicago sativa*), barley (*Hordeum vulgare*), maize, soybean, and wheat (USGCRP, 2014).

- Due to favorable weather conditions, soil fertility, and topographic characteristics (i.e., almost flat topography), maize is intensively cultivated in Iowa, Midwestern USA (Green et al., 2018). The state is consistently the top producer of maize in the United States (EDF, 2022). However, due to the strong dependency of maize yield on rainfall (i.e., rainfed fields), climate change aspects, especially rainfall variability, have resulted in substantial changes in maize yields across the state over recent years (Joshi et al., 2021). Over the next few decades, maize yields are likely to experience a climate burden in many regions of the state (EDF, 2022). For instance, driven by increases in extreme heat, the expected climate will decrease maize yields in Davis County is 44% by 2050 (EDF, 2022). Overall, almost all counties in Iowa will face maize yields that are more than 5% lower than they would have been without climate change (EDF, 2022).

- In subarctic Alaska, with appropriate adaptation to future climate conditions (e.g., improved cultivars), climate change could provide an opportunity to increase yields of some crops and enhance food security (e.g., Harvey et al., 2021). For instance, an increase in the length of the growing season and the associated increase in the number of growing degree days could provide an opportunity to enhance growth conditions of spring wheat and increase its yields (ACIA, 2005; Harvey et al., 2021). Harvey et al. (2021) is one of the few studies that made projections of the effect of climate change on Alaska's wheat crop yield. Specifically, the study showed that improved cultivars could enhance wheat yield by 5% and 7% during 2020–2049 under medium (RCP4.5) and high (RCP8.5) GHG emission scenarios (relative to the 1989–2018 baseline period). Relative to the same reference period (1989–2018), the study also revealed that with improved cultivars and soil water at 85% field capacity at planting, wheat yield could be increased by 11% and 12% in 2035 under RCP4.5 and RCP8.5 scenarios, respectively. This study shows that appropriate agronomic practices (e.g., suitable selection of crop varieties, good soil management, planting date modification, etc.) are of great importance to cope with climate change in the agriculture sector.

- In the Texas High Plains (THP), maize is one of the dominant crops and farmers typically select long-season maize varieties to grow under irrigation (e.g., Xue et al., 2017; Schnell et al., 2021). Annually, 53% of the total agricultural water resource budget in the THP is used to irrigate maize (Xue et al., 2017). Climate change by 2050 in the THP and its associated decrease in growing season length are anticipated to affect the yield of irrigated maize (e.g., Kothari et al., 2022). For instance, Kothari et al. (2022) projected climate change effects on grain maize yield at Bushland in the Northern High Plains of Texas using the CERES-Maize model and found that grain maize yield under RCP4.5 and RCP8.5 scenarios is expected to decrease by approximately 31% by 2050 (relative to the 1976–2005 baseline period). Potential shorter crop season and decreased unit grain weight and biomass are expected to be the main causes for this yield reduction (Kothari et al., 2022). The same study also revealed that late planting (beyond the reference planting date of mid-May) can result in an increase in grain yield and a decrease in seasonal irrigation water requirement under climate change by 2050.

- In California, one of the USA's most agriculturally productive regions, droughts are expected to adversely affect crop production by slowing plant growth and resulting in critical crop yield losses by mid-century (Pathak et al., 2018). As agriculture in this region is highly dependent on irrigation, droughts are expected to lead to a decrease in water availability (Dogan, 2015), potentially reducing irrigated crop areas and yields and altering crop irrigation requirements (Medellín-Azuara et al., 2018). For instance, relative to today, the South Coast region in California (which hosts most of the state's population) is projected to lose up to 22% of its current irrigated area due to climate change by 2050 (Medellín-Azuara et al., 2018). Additionally, potential warmer winters by 2050 in California are projected to lead to a decrease in wheat area and an increase in alfalfa and tomato area (Lee and Sumner, 2015). Under climate change by 2050, yields of crops such as cucurbits, alfalfa, rice, almonds, maize, wheat, cotton, tomato, onion, garlic, potato, and sugar beet are expected to show changes within ±10% of standard values (Medellín-Azuara et al., 2018). For instance, alfalfa yields could be enhanced (e.g., Lee et al. 2011), while climate change would reduce yields of cucurbits (e.g., Kerr et al. 2018). These results demonstrate that each crop responds differently to climate change and effective adaptation measures should be developed based on the specificity of each crop.

- The United States is the world leader in soybean production, contributing more than 30% of global production in 2014 (Timmerman et al., 2023). This economically crucial crop is largely produced in the Southeastern part of the country (Bao et al., 2015) and plays a fundamental role in ensuring food security and biofuels production (e.g., Pimentel and Burgess, 2014). Previous studies assessing the future effects of climate change on soybean yield have produced mixed findings, with both positive (e.g., Bao et al., 2015) and negative (e.g., Schauberger et al., 2017) implications of increases in temperature and changes in rainfall patterns on soybean yields. For instance, Bao et al. (2015) projected that relative to the 1923–2008 baseline period, potential increases in precipitation over the soybean growing season and CO_2 concentration may lead to an increase in rainfed soybean yields by 8 to 35% by 2050. The same study also revealed that projected increases for irrigated soybean yield were approximately 1 to 12% less than for rainfed soybeans. However, another study conducted by Schauberger et al. (2017) estimated that soybean yields over the coming decades are projected to decrease at temperatures higher than 30°C and that potential increases in CO_2 concentration can only weakly reduce these yield losses, in contrast to irrigation. These contradictions could be due to varying degrees and directions of changing climate parameters linked with distinct geographical regions studied, uncertainties associated with the models used, and model assumptions (Su et al., 2021; Sharma et al., 2022). Therefore, further local studies are highly recommended to decrease uncertainties associated with soybean yield projections.

According to Wang et al. (2015), many of the regions mentioned above are among the world's top crop-producing areas and are at the forefront of international efforts to address global food security. Projections of crop yields indicate that, in the absence of adequate intervention, climate change could impede crop yield growth in these regions. This may threaten the stability of food supply, access, and utilization in many countries, particularly developing countries that rely heavily on food imports from the United States (Clapp, 2017).

10.3   CROP YIELDS BY 2050 ACROSS CANADA

Climate change in Canada over the past few decades has had both positive and negative impacts on crop production (Qian et al., 2019). As a high-latitude country, Canada is likely to experience less severe warming in the coming decades than lower-latitude countries (Lobell et al., 2011). With adequate precipitation, some previously unproductive northern areas with fertile soils may become productive (Simon, 2022). By 2050, Canada can expect warmer temperatures, significant changes in precipitation patterns, and more frequent extreme weather events such as droughts and heatwaves (e.g., Li et al., 2018). These changes could significantly affect crop yields over time, making it crucial to understand how Canada's crop yields may change under climate change by 2050. This is particularly important for adapting to anticipated changes in climate conditions. Accordingly, many previous studies at the regional scale have used various methodologies and models to project Canadian crop yields under different climate change scenarios by 2050 (e.g., Carew et al., 2017; He et al., 2018; Adekanmbi et al., 2023). This section summarizes these regional-scale projections (Figure 10.2).

- In Manitoba, which accounts for up to 17% of Canada's total annual wheat production (Census of Agriculture, 2011), average temperatures are expected to rise by 1–3°C by 2050, with more heat, fewer cold spells, and longer growing seasons (Blair et al., 2016). As a result, wheat yields in the region are projected to increase by 2050 (Laforge et al., 2021). For example,

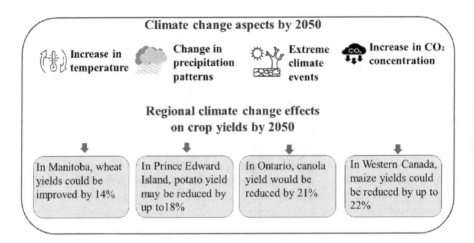

FIGURE 10.2   Effects of climate change on Canadian crop yields by 2050.

Carew et al. (2017) estimated that, relative to the 1996–2012 baseline period, climate change in the region over 2034–2050 would increase average wheat yields by up to 14% under high-carbon scenarios. This potential increase in wheat yield could be attributed to the strong effect of CO_2 (Laforge et al., 2021). Since CO_2 is a key element in photosynthesis in crops, an increase in atmospheric CO_2 may contribute to improved yields through increased photosynthesis, known as the "CO_2 fertilization effect" (Laforge et al., 2021). This effect is stronger in some crops such as wheat (Xie et al., 2020; Laforge et al., 2021).

- In Western and Eastern Canada, He et al. (2018) used the Canadian Regional Climate Model (CanRCM4) and the DSSAT model to show that, relative to the baseline climate scenario (1971–2000), climate change would increase wheat yields by 8% and 11% by 2050 under medium (RCP4.5) and high (RCP8.5) GHG emissions scenarios, respectively. However, by 2050, maize yields are expected to decrease by 15% and 22% under both scenarios, respectively (He et al., 2018).
- In Southern Canada, climate change is projected to increase the length of the growing season for warm-season crops by about 20 days by 2050 (Li et al., 2018), providing farmers with an opportunity to grow new crop varieties (e.g., new wheat and maize cultivars) that may exhibit higher yields under climate change (He et al., 2018).
- In Prince Edward Island (PEI), where high-quality potatoes are an important part of Canada's agricultural exports (Agriculture and Agri-Food Canada, 2021), climate change over the coming decades may have negative impacts on yields of this rainfed crop. For example, Adekanmbi et al. (2023) found that under a high GHG emissions scenario (e.g., SSP5-8.5), potato yields in PEI could decline by about 18.8% by 2050 compared to the 1995–2014 baseline period. This highlights the urgency of developing effective climate adaptation options (e.g., supplemental irrigation, efficient

soil management) to ensure the long-term sustainability of potato production in the region (Adekanmbi et al., 2023).

- In Ontario, warmer climate change by 2050 may reduce canola (*Brassica napus* L.) yields due to the crop's rapid response to heat stress (Morrison and Stewart, 2002). Canola is a fundamental oilseed crop in Ontario and Canada that contributes significantly to Canada's agricultural exports (FAOSTAT, 2015). Overall, Canada exports about 90% of its canola production (FAOSTAT, 2015). Qian et al. (2018) found that, relative to the 1971–2000 baseline period, canola seed yield would decrease by 21% in 2050 under a medium (RCP4.5) GHG emissions scenario. The study also showed that this projected seed yield reduction would be accompanied by increases in heat and water stress (rainfall shortage) during the canola growing season. Therefore, coping with heat and water stress is essential for ensuring sustainable canola production in the region and in Canada under climate change by 2050 (Qian et al., 2018).

- In Northern Southwestern Ontario, Simon (2022) projected yields of irrigated soybeans under a high (RCP8.5) GHG emissions scenario over the next few decades and found that with adequate moisture, soybean yields are expected to continue increasing over time as a result of warming temperatures and some inherent CO_2 fertilization until 2050. The same study also found that rising temperatures over the coming decades would pose the greatest challenge to high maize productivity in Southwestern Ontario. Specifically, reduced precipitation would be a major concern for changes in soybean yield while rising temperatures would be a major concern for changes in maize yield (Simon, 2022).

Based on the above projections of Canadian crop yields by 2050, it is clear that climate change in Canada could present both challenges and opportunities for enhancing agricultural production. In both cases, key adaptation measures to boost crop yields in the coming years are essential to meet rising food demands. By 2050, Canadian food production must increase by 25% just to maintain Canada's contribution as the global population grows (RBC, 2022). Boosting crop yields to meet these rising demands, rather than clearing more land for agriculture, has been recommended as a good option for achieving this goal (Ray et al., 2013). Key adaptation measures are those that ensure sufficient food production while reducing GHG emissions.

10.4   CONCLUSIONS

Numerous previous studies have combined climate, crop, and economic models to assess the impact of climate change on crop production across the United States and Canada. These studies have shown that while some regions in the United States and Canada may experience favorable conditions for crop growth as a direct result of warmer climate by 2050, in most regions, aspects of climate change such as droughts, rainfall variability, and floods are expected to slow crop growth by that year, even under an optimistic climate scenario. Much worse climate scenarios by 2050 could lead to unprecedented declines in crop yields. Due to the high dependence of many countries on agricultural exports from the United States and Canada to feed their populations, it is urgently necessary to accelerate adaptation efforts in both countries to protect global food supplies. Adaptation measures in both countries can ensure that farms remain productive in the face of climate change and continue to support global food demands. It is worth noting that implementing adequate adaptation measures takes time to work successfully, so decision-makers should act as soon as possible to avoid the worst global food insecurity scenarios by 2050. If adaptation measures are successfully implemented, the United States and Canada could significantly increase their agricultural production and grow their exports and output by 2050.

Part III

Adaptation Strategies to Address Climate Change for Global Food Security

To feed the world's population of 9.4 billion by 2050, we must ensure that our agriculture continues to produce enough food under climate change. We need to identify the most effective options of adaptation in the agriculture sector that must be implemented over the next few years. Sustainable agriculture is emerging as the only option as it helps to increase agricultural production and preserve natural resources by reducing greenhouse gas emissions, sequestering more carbon in the ground, conserving water and energy, and protecting biodiversity and natural habitats. As some past and current agricultural practices are already under question environmentally, the issue of climate change and its effects on agricultural production increasingly becomes more complex. Can we successfully and sustainably ensure enough global agricultural production by 2050 for about 9.4 billion people? One of the fundamental ways which can accelerate the successful transition to agricultural production security by 2050 is leveraging research and innovation and making output of science practical. Chapters 10–15 of this book discuss appropriate measures and policies that can be helpful for achieving sustainable agriculture and food security by 2050 and key considerations for their implementation. The time to act on adaptation in agriculture and food security is now.

11 Options and Constraints for Meeting the Challenge of Increasing Global Agricultural Production

11.1 INTRODUCTION

Under projected climate change conditions in 2050 (e.g., increased temperature and rainfall variability), global agricultural production must increase by up to 56% relative to 2010 levels to meet the food demands of an additional 2.5 billion people (United Nations, 2019; van Dijk et al., 2021). Specifically, the production of major global staple crops (e.g., wheat, barley, maize, soybean, and rice), livestock products, and fisheries must increase effectively by 2050 (van Dijk et al., 2021). This presents a major challenge for humanity, particularly in light of expected climate change over the next few decades (Howden et al., 2007). Despite uncertainties about future climate scenarios and their potential effects, numerous studies have revealed that climate change will reduce overall agricultural productivity in the coming decades (Arora, 2019; Malhi et al., 2021; Bai et al., 2022). These studies also indicate that agriculture must become more efficient, resource-saving, and environmentally friendly to adapt to climate change. Fortunately, many adaptation options have the potential to reverse the negative effects of climate change (Malhi et al., 2021). However, due to differences in climate and other environmental variables, cultural, institutional, and economic factors; and their interactions, there is a large array of possible adaptation options (Howden et al., 2007). Broadly speaking, there are two major options for increasing global agricultural production: (1) enhancing yields per unit area of current croplands through sustainable methods aligned with the Food and Agriculture Organization's vision of Sustainable Food and Agriculture goals; and (2) reforming agricultural practices to mitigate greenhouse gas emissions from agriculture (Zilberman et al., 2018) (Figure 11.1). However, constraints such as misunderstanding by farmers and nonalignment of adaptation measures with specific conditions of agricultural fields may reduce the performance of these options (Howden et al., 2007). Addressing these constraints is necessary for more effective adaptation. The objectives of this chapter are to outline available adaptation options for increasing global agricultural production under climate change and to summarize the main limitations and challenges that may reduce the performance of these options.

DOI: 10.1201/9781003404194-14

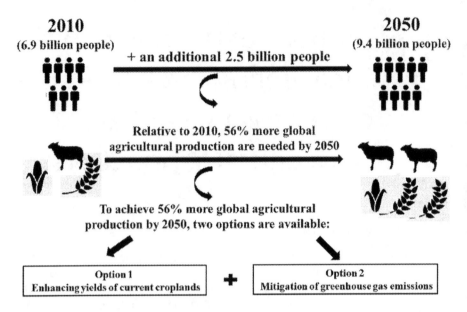

FIGURE 11.1 Necessary measures to enhance global agricultural production and to achieve sustainable agriculture by 2050.

11.2 OPTION 1: ENHANCING THE YIELDS OF CURRENT CROPLANDS

Recent years have seen global agricultural production grow at a rate of about 1% per year, much lower than the world's population growth rate (Hemathilake and Gunathilake, 2022). This weak agricultural growth, combined with the progressive increase in global population and potential decrease in crop yields in the coming decades, has led the international community to explore new scientific and cultivation options to achieve higher productivity by 2050 (Acevedo et al., 2018). New adaptation options have been developed to adjust crop requirements to new climate realities and reduce the risks of climate change (Ahmad et al., 2020). The following subsections outline complementary recent adaptation measures that can increase global agricultural production under climate change without damaging the environment (Figure 11.2).

11.2.1 DEVELOPMENT OF IMPROVED CROP VARIETIES

Developing improved crop varieties is an effective, easy, and economical adaptation option for sustaining crop production under climate change (Ahmed et al., 2019). This option involves developing new crop varieties such as early maturing, drought and heat tolerant varieties that are suited to new climate conditions (Simtowe et al., 2019). Improved crop varieties have the potential to increase production per unit area under various environmental stresses such as salinity, moisture, and extreme climate events (Deressa et al., 2009). They also typically exhibit high disease resistance, good responses to fertilizers, and high-quality products at the end of the growing

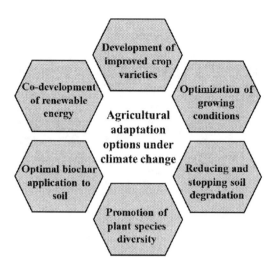

FIGURE 11.2 Recent adaptation measures that can increase global agricultural production without environmental deterioration.

season (Simtowe et al., 2019). Crop variety tests are necessary to identify new varieties with high yield, quality, yield stability, disease resistance, and wide adaptability (Pan et al., 2022). Recently, with the widespread use of artificial intelligence technology and machine learning, a range of variety test platforms such as BMS (Integrated Breeding Platform, 2019), AGROBASE (Agronomix Software, 2022), BIMS (Jung et al., 2021), Genovix (Agronomix Software, 2022), PRISM (Central Software Solutions, 2022), and GoldenSeed (Han et al., 2018; Zhao et al., 2022) have been developed to help users (e.g., researchers, farmers, decision-makers) conduct crop variety tests efficiently through functions such as test design, data processing, and statistical analyses (Yang et al., 2023a,b). With standardized data from these platforms, the construction of variety knowledge maps and intelligent analysis models combined with climate data will enable the development of intelligent decision support systems for variety promotion and improvement of crop management under climate change (Yang et al., 2023a,b).

Recent advancements in improved crop varieties (e.g., wheat, maize, barley) have been discussed in numerous studies, and several improved varieties have been identified as playing a key role in climate change tolerance (Toulotte et al., 2022). For example, in Switzerland, Friedli et al. (2019) evaluated 14 bread wheat (*Triticum aestivum* L.) genotypes spanning the past hundred years of Swiss wheat breeding progress and concluded that modern wheat varieties have adjusted rooting depth to water demand, enhancing adaptation to drought events as crop roots can move in search of water (Hawes et al., 2000). In Bangladesh, where rice is crucial for food security, Bairagi et al. (2021) revealed that submergence-tolerant rice varieties (BRRI dhan51, BRRI dhan52, BRRI dhan79, BINA Dhan 11, and BINA Dhan 12) developed by the Bangladesh Rice Research Institute and the Bangladesh Institute of Nuclear Agriculture are suitable for increasing yields in flood-prone areas. However, further efforts are needed to encourage farmers to adopt these improved

TABLE 11.1
Examples of Recently Identified Crop Varieties that Show Promise for Increasing Agricultural Production in the Face of Climate Change

Region	Improved Crop Variety	Specific Region/ Country	Reference
Europe	Buckwheat (*Fagopyrum esculentum Moench*)	The Mediterranean	Kakabouki et al. (2021)
	Spelt wheat (*Triticum spelta* L.)	The Mediterranean	Kakabouki et al. (2021)
	Sweet potato (*Ipomoea batatas Lam.*)	The Mediterranean	Kakabouki et al. (2021)
	Barley (*Hordeum vulgare*) (*IPZ 24727*)	Germany	Hu et al. (2021)
	Wheat (*T. aestivum L. cv. Alka*)	Turkey	Yoruk et al. (2018)
Africa	Maize (*Zea mays* L.) (*BH-540*)	Ethiopia	Abate et al. (2022)
	Soybean (*Glycine max* L. Merril) (*TGX1951-3F*)	West Africa	Bebeley et al. (2022)
	Maize (*Zea mays* L.) (*CZH0616*)	Eastern and southern Africa	Setimela et al. (2017)
Asia	Rice (*Oryza sativa* L.) (*BRRI dhan79*)	Bangladesh	Bairagi et al. (2021)
	Potato (*Solanum tuberosum*) (*Kufri Pukhraj*)	India	Pradel et al. (2019)
	Rice (*Oryza sativa* L.) (*KD-SKC1-FL*)	Thailand	Punyawaew et al. (2016)
Americas	Wheat (*TAM112*)	The United States	Reddy et al. (2014)
	Wheat (*Pelissier*)	Canada	Ashe et al. (2017)

varieties. In eastern and southern Africa, where maize (*Zea mays* L.) is the main source of food security and economic development, new drought-tolerant maize hybrids and open-pollinated varieties (e.g., CZH0616 variety) have been identified in recent years that exhibit strong resistance to droughts and yield up to 20% more than older varieties (Setimela et al., 2017). In West Africa, where soybean (*Glycine max* (L.) Merril) production has been increasing in recent years (MacCarthy et al., 2022), new soybean varieties have been identified and recommended for optimum grain yield in light of projected climate change in the coming decades (Bebeley et al., 2022). More examples of improved crop varieties identified in recent years throughout the world are summarized in Table 11.1.

11.2.2 OPTIMIZATION OF GROWING CONDITIONS

Optimization of growing conditions involves identifying optimal growth conditions (e.g., planting time, spacing, crop migration, planting density) that can enhance crop growth and yield under different climate and environmental conditions (Xu et al., 2021). Many studies have been conducted to identify these optimal growth conditions. For example, Traore et al. (2017) found that early planting coupled with an optimal mineral fertilizer rate can buffer maize yield loss in southern Mali. Ren et al. (2016) found that increasing planting density is a good option for enhancing maize yield in the Loess Plateau of China, as higher plant populations increase the number of maize ears (Xu et al., 2021). In the North China Plain, Sun et al. (2016) revealed that keeping the sowing date within a certain range after June can improve summer maize yield. More recently, Xu et al. (2021) used the APSIM-Maize model to analyze maize yield changes under 20 different combinations of sowing dates and planting densities and found that delaying sowing time by 1–2 weeks and increasing planting density to 8–10 plants m^{-2} can reduce spring maize yield losses by about 27% and 45% under medium (RCP4.5) and high (RCP8.5) GHG emission scenarios by 2050, respectively (relative to the 1971–2010 baseline period). Crop migration is also emerging as an effective climate change adaptation option due to its contribution to crop yield improvement (Sloat et al., 2020; Ologeh et al., 2021). Many aspects of climate change such as droughts and rainfall variability are expected to profoundly alter growing conditions for agricultural crops, potentially resulting in reduced crop yields in many agricultural areas worldwide (Skarbø and VanderMolen, 2016). A switch from agricultural areas currently experiencing progressive decreases in crop yields due to increasingly higher temperatures and more frequent extreme climate events to new areas better adapted to new climate conditions has been proposed as one possible adaptation option, referred to as crop migration (Sloat et al., 2020). In these new areas, increased atmospheric carbon dioxide (CO_2) concentration and warmer climate can promote photosynthesis, enhance crop yield, and promote the shift of crop species to areas previously constrained by either too short a growing season or unreachable thermal requirements for completing the crop growth cycle (Marx et al., 2017; Ceglar et al., 2019). Overall, Sloat et al. (2020) suggested that the most damaging effects of climate change on yields of major rainfed crops (mainly maize, wheat, and rice) can be moderated by crop migration over time and expansion of irrigation.

However, factors such as migration mode, migration destinations, land suitability, climate conditions, and socio-economic and political factors of the destination location must be carefully assessed to avoid environmental issues (Sloat et al., 2020). Crop migration at a local level has gained attention in recent years due to its potential to produce higher yields and high-quality products (Arias et al., 2022). For example, grape cultivation that has migrated to high-altitude areas in Argentina, Brazil, China, Italy, Portugal, and Turkey in recent years has shown notable enhancements in crop yields (Arias et al., 2022). In Eastern Europe, crop migration is expected to reach twice the velocity observed during the period 1975–2016 over the next few decades (Ceglar et al., 2019). Many Mediterranean European countries may lose suitability for growing specific crops in favor of northern European countries (Ceglar et al., 2019). Increased growing season length in northern European countries allows for expansion of agriculture, introduction of crops historically cultivated in warmer regions, and crop diversification, contributing to global food security (Wiréhn, 2018; Unc et al., 2021). It is important to note that sustainable development of agriculture in new growing areas requires local solutions supported by locally relevant policies (Unc et al., 2021).

11.2.3   Reducing and Stopping Soil Degradation

Agricultural soil degradation, the loss of soil's production capacity mainly due to erosion and salinity processes, is a growing issue worldwide that critically affects agricultural production and food security (Pimentel and Burgess, 2013). Annually, the world loses 75 billion tons of soil, resulting in an economic loss of about $18–20 trillion (Pimentel and Burgess, 2013; UNCCD, 2019; Gobinath et al., 2022). About 30% of the world's soils have been exhausted and require improved management to enhance agricultural production (Nascimento et al., 2021). Without proper soil management, loss of soil organic carbon due to soil degradation may reach 212 Gt by 2050, affecting the ability of global agriculture to provide food (UNCCD, 2019). Due to these alarming rates of degradation globally, urgent conservation actions are needed to stop and reverse declines in soil degradation and adverse soil quality in light of mounting global demands for agricultural products (Hossain et al., 2020). This is crucial for ensuring the long-term productivity of agriculture. Reducing and stopping agricultural soil degradation involves raising awareness of the issue and encouraging innovative approaches to soil management under climate change (UNCCD, 2019). The following subsections provide up-to-date information on how to enhance the productivity of degraded soils under climate change.

- **Enhancing productivity of eroded soils under climate change**
  Soil erosion, the loss of topsoil via agents such as water, wind, and tillage, is a global issue that results in the loss of about 10 million ha of cropland annually (Pimentel and Burgess, 2013), leading to an economic loss estimated at \$40–490 billion (Nkonya et al., 2016) and a reduction in global agri-food production by about 33.7 million tons yearly (Sartori et al., 2019). Recent estimates indicate that about 1.09 billion ha of global land has been affected by water-driven soil erosion (Lal, 2017). According to the Global Soil Partnership of the FAO, it is estimated that 75 billion tons of soil are eroded every year from arable lands worldwide (Borrelli et al., 2017). Soil erosion may also result in off-site effects such as sedimentation, flooding, damage to infrastructure, landslides, and water eutrophication (Boardman and Poesen, 2006; Sartori et al., 2019). Using the latest projections of climate and land use change, Borrelli et al. (2020) found that relative to 2015, climate change impacting land use may increase the global soil erosion rate by 2%–10% by 2070, depending on the climate change scenario. This can be explained by the fact that soil erosion under climate change is most directly affected by changes in extreme precipitation (Nearing et al., 2004; Eekhout and Vente, 2022), which is expected to increase due to the increasing moisture-holding capacity of a warmer atmosphere, leading to more aggressive precipitation events and more intense soil erosion (Trenberth, 2011; Eekhout and Vente, 2022). However, if soil conservation measures are taken, it is possible to entirely reverse the effect of climate change (Eekhout and Vente, 2022). Therefore, urgent soil management steps must be taken to address soil erosion (Gobinath et al., 2022). Measures such as reduced tillage, contour farming, terraces, afforestation of slopes, plant residues, cover crops, grass margins and brush layers are often recommended by researchers as the best options for decreasing soil erosion (Panagos et al., 2016; Poesen, 2018; Sartori et al., 2019; Eekhout and Vente, 2022). Overall, Borrelli et al. (2020) showed that these measures can decrease the global soil erosion rate by 5% under climate change by 2070. In addition to their key role in reducing soil erosion, these measures also have great potential to increase carbon sequestration and enhance biological activity, aggregate stability, and overall soil health (Eekhout and Vente, 2022). They can also

help achieve UN Sustainable Development Goals and UN strategies for soil conservation. Understanding suitable adaptation measures against soil erosion now and in the future is crucial for addressing climate change issues (UNCCD, 2019).

- **Enhancing productivity of saline soils under climate change:**
  Soil salinity, characterized by a high concentration of salts in the soil, can negatively impact crop mechanisms and enzymatic activities, leading to a significant decline in crop yields and agricultural productivity (Haj-Amor et al., 2022). With approximately 19% of global land area (935 million ha) affected by soil salinity (Ivuskin et al., 2019) and an annual decrease of 2% (10 million ha) in the world's agricultural areas due to soil salinity (IPCC, 2019), mitigating salinity and enhancing the productivity of saline soils is crucial for ensuring global food security in the coming decades. Traditional reclamation measures such as leaching salts from the root zone via drainage, scraping, surface flushing, tillage, fertilization, gypsum addition, and promotion of salt-tolerant crops have been suggested as effective options for reducing soil salinity levels and increasing productivity (Keren, 2005; Gill et al., 2009). However, in light of the strong negative impact of climate change on soil salinity, innovative recent options such as biochar use, co-addition of biochar and microorganisms, and use of salinity-resistant microorganisms from the rhizosphere of halophyte plants have emerged as the most sustainable measures for enhancing productivity under various aspects of climate change such as droughts, rainfall variability, and heatwaves (Haj-Amor et al., 2022).

11.3 OPTION 2: REDUCING GHGS EMISSIONS THROUGH LAND RESTORATION

Due to climate change and soil degradation, global agricultural production is projected to decrease by 10%, with some countries (especially in Africa) experiencing decreases of up to 50% (UNCCD, 2019). To mitigate these alarming projections, there is a pressing need to reduce GHG emissions from degraded soils (UNCCD, 2019). Soils play a crucial role in climate mitigation as both a carbon sink and a source of GHG emissions (UNCCD, 2019). For example, between 2000 and 2009, soil degradation was responsible for an estimated 3.6–4.4 billion tons of CO_2 emissions (UNCCD, 2019). However, restoration of degraded lands can play a significant role in mitigating GHG emissions and advancing global food security (Toensmeier, 2016). Specifically, land restoration, particularly the restoration of abandoned agricultural lands, has the potential to decrease carbon emissions by capturing and storing approximately 1–3 Gt of CO_2 annually (UNCCD, 2019) while also enhancing ecosystem services such as biodiversity, water purification, and erosion control (Yang et al., 2019a,b,c). Agricultural land restoration can be achieved through traditional agronomic and biological methods such as crop rotations, agroforestry, integration of livestock into cropping systems, reduced tillage, cover crops, and compost application (Saturday, 2018). The benefits of land restoration are on average ten times greater than the total costs of inaction (Nkonya et al., 2016; UNCCD, 2019).

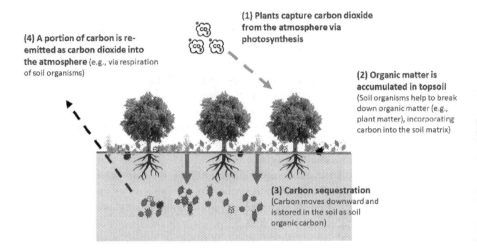

FIGURE 11.3   Steps of carbon sequestration in biomass and soil.

Abandoned agricultural lands are often considered the most effective for capturing and storing atmospheric CO_2 (Yang et al., 2019a,b,c). Globally, approximately 430 million hectares of land have been cropped, degraded, and then abandoned – particularly in high- and middle-income countries (Campbell et al., 2008; Yang et al., 2019a,b,c). Once degraded lands are abandoned and undergo ecological succession, they remove atmospheric CO_2 and sequester its carbon as soil organic matter through the process of carbon sequestration (Lal, 2004; Yang et al., 2019a,b,c) (Figure 11.3).

The process of carbon sequestration is typically slow, requiring approximately 100 years or more to re-attain pre-agricultural soil C levels (Yang et al., 2019a,b,c). To address this constraint, scientists have recently identified methods to accelerate the carbon sequestration process. This has led to the launch of the Group on Earth Observations (GEO) Land Degradation Neutrality flagship initiative by 114 governments and 144 participating organizations of the GEO, with the goal of accelerating land restoration through improved methods (UNCCD, 2022). The Flagship supports the UN Sustainable Development Goal (SDG) to accelerate the ecological restoration of approximately 1 billion ha of land by 2030 (UNCCD, 2022). The following subsections summarize key strategies for accelerating the C sequestration process as suggested by Yang et al. (2020).

11.3.1   Promotion of Plant Species Diversity

Incorporating a greater variety of plant species and effectively managing high plant diversity in degraded agricultural lands can significantly enhance soil organic carbon sequestration and promote total C storage over a relatively short period of time (Yang et al., 2020). For example, Wang et al. (2020) demonstrated that in the Loess Plateau, China, rates of soil organic C sequestration on restored grasslands, shrublands, or forests were 92 to 215% higher than those under natural recovery. Similarly, Manaye et al. (2019) found positive effects of woody species diversity, abundance, and richness on C storage in degraded agricultural lands in Ethiopia. Specifically, they found that increasing the number of woody species used in restoration from 26 to 46 could increase total C storage by up to 197%. Dayamba et al. (2016) also reported a positive relationship between plant species diversity and C storage in Western Africa. However, it is important to carefully assess and identify the characteristics of local plant species and plant diversity to avoid any negative effects of land restoration projects (Manaye et al., 2019). Additionally, strategies to increase plant species diversity should be facilitated by governmental and corporate initiatives at regional levels (Yang et al., 2020).

11.3.2 BIOCHAR APPLICATION TO SOIL

Biochar is a high-carbon, fine-grained residue produced through the thermochemical conversion of various types of biomass in an oxygen-limited environment (Gross et al., 2021). It is primarily applied to soil to improve soil health and plant productivity (Gross et al., 2021). In addition to its positive effects on nutrient availability, soil-water content, microbial biomass, soil microbial diversity, and crop yields, biochar application can also sequester carbon in the long term due to its high stability in soil (Gross et al., 2021). Furthermore, biochar can reduce agricultural greenhouse gas emissions, particularly nitrous oxide (N_2O) and methane (CH_4). Specifically, it can significantly decrease N_2O emissions (Borchard et al., 2019), suppress CH_4 release in flooded and/or acidic soils (Jeffery et al., 2016), and enhance native soil organic matter (Wang et al., 2016; Yang et al., 2019a,b,c). However, the reactions of agricultural GHGs with biochar addition are context-specific. Therefore, biochar should only be applied to abandoned land if the specific conditions of the land (soil properties, climate conditions, and management practices) are favorable for such an addition (Yang et al., 2019a,b,c).

11.3.3 CO-DEVELOPMENT OF RENEWABLE ENERGY

This option involves the co-development of renewable energy and carbon sequestration initiatives on abandoned agricultural lands (Yang et al., 2019a,b,c). The establishment of renewable energy technologies on abandoned agricultural lands is becoming increasingly attractive as a means of creating new development opportunities in the agricultural sector worldwide (Bell et al., 2022). For example, Tumelienė et al. (2022) assessed the suitability of abandoned agricultural lands covered with woody plants for wind energy production in Lithuania and found that 7% of these lands could be profitable for both landowners and energy developers, potentially contributing to the country's growing energy consumption. However, it is important to avoid establishing renewable energy technologies on productive agricultural lands to prevent the loss of agriculturally productive land and concerns about food security (Yang et al., 2019a,b,c; Tumelienė et al., 2022).

11.4   CHALLENGES AND LIMITATIONS

Despite the promising potential of climate change adaptation options to mitigate the negative effects of climate change on agricultural productivity and land degradation by 2050, as discussed in Sections 2 and 3, there are still challenges and limitations that may hinder the successful implementation of some adaptation measures. For example, while the effects of land restoration measures on soil functions, food production, and carbon sequestration in soil are well known, little is known about their potential effects on human communities. Future research should focus on addressing this knowledge gap to avoid any unexpected negative impacts (Newton et al., 2021). Additionally, stronger linkages between proposed adaptation measures and Sustainable Development Goals (SDGs) such as SDG 2 (achieving food security and promoting sustainable agriculture) and SDG 13 (taking action on climate change) are needed (Hamidov et al., 2018). Furthermore, as agricultural land abandonment is projected to continue in many regions such as Europe, Russia, Central Asia, East Asia, and the Americas over the next few decades (Popp et al., 2017; Sanderson et al., 2018; Leclère et al., 2020; Crawford et al., 2022), accurate information about where and when abandonment will occur would be helpful in mitigating some of the anticipated effects of climate change on the agricultural sector (Crawford et al., 2022).

It is important to note that many internal factors, such as the age of farmers, education level, household size, income, access to agricultural extension services, credit, and information, also play a key role in ensuring the successful implementation of adaptation measures (Dang et al., 2019). However, little research has been conducted to understand how these internal factors affect farmers' adaptation measures, indicating a need for further investigation in this area (Dang et al., 2019; Saddique et al., 2022). Additionally, decision-makers often only consider agricultural adaptation measures published in English while relevant studies written in other languages (particularly local ones) are ignored. This can result in the omission of potentially helpful adaptation measures (Jiang et al., 2023). In light of the ongoing impact of climate change on agriculture worldwide, it is crucial for international decision-makers to be informed of all relevant information, regardless of the language in which it is written. Communication through social media, television, newspapers, and other channels plays a key role in raising awareness and motivating decision-makers and the public to take action to mitigate the effects of climate change on agriculture (Depoux et al., 2017). Finally, it is worth noting that while adaptation measures such as soil-based carbon sequestration and soil conservation actions can significantly help combat climate change, they cannot remove carbon from the atmosphere as quickly as it is being added by energy production, industrial activities, and agriculture. To avoid critical global warming by 2050 (i.e., global temperature increase >1.5°C), efforts to store carbon must be coupled with drastic reductions in greenhouse gas emissions (Melillo et al., 2017).

11.5   CONCLUSIONS

This chapter reviews the available adaptation options for increasing global agricultural production under climate change by 2050, along with their limitations and challenges. The review process revealed that traditional agricultural adaptation measures such as irrigation and drainage, tillage, fertilization, and gypsum application should be supplemented with more recent measures such as biochar application to soil and the co-development of renewable energy and carbon sequestration initiatives. These measures can increase yields on farmland, particularly degraded land, and also lead to environmental improvements by reducing greenhouse gas emissions such as CO_2, CH_4, and N_2O from agricultural activities. The successful implementation of these measures in the coming years is crucial for achieving both sufficient agricultural production for a growing global population and environmental protection. However, more effort is needed to introduce developed adaptation measures to decision makers for rapid funding and planning actions, as well as to educate farmers on the benefits of available agricultural adaptation measures.

12 European Policies and Measures for Aligning Climate Targets with Agricultural Productivity Goals

12.1 INTRODUCTION

The observed climate change and warming trend are evident in all European countries (e.g., Kovats et al., 2011; Ionita and Nagavciuc, 2021) and are projected to continue increasing between 2023 and 2050 at a rate similar to past observations, regardless of future GHG emissions scenarios (Copernicus Climate Change Service, 2022). This could have critical effects on agricultural productivity (Hristov et al., 2020). If no adaptation measures are implemented, projected changes in daily temperature, precipitation, wind, relative humidity, and global radiation relative to the baseline period of 1981–2010 may reduce yields of major European crops such as maize and wheat by up to 49% and 22%, respectively, by 2050 (Hristov et al., 2020). Furthermore, some European regions may experience declines in yields of rainfed crops of up to 50% during 2021–2050 relative to the same baseline period (Verschuuren, 2022).

In light of these projections for 2050, there is currently great potential in European Union (EU) countries to increase agricultural productivity in the coming decades to nutritiously and sustainably feed the growing European and global populations in the context of climate change (Candel and Biesbroek, 2018). These projections have also resulted in calls for better integrated measures and policies to govern agricultural production and food security (Candel and Biesbroek, 2018). Currently, EU countries are implementing a wide array of short- and medium-term measures (e.g., introduction of new cultivars, improved irrigation, better crop practices) and policies such as the Common Agricultural Policy (CAP), the Common Fisheries Policy (CFP), and biofuels targets to strengthen food systems, enhance soil and water resource management at the farm level, increase agricultural productivity, and support farmers and consumers in the face of climate change (Candel and Biesbroek, 2018; Zhao et al., 2022).

In addition to the challenge of ensuring sufficient agricultural production by 2050, EU countries must also meet the 1.5°C climate target by 2050. This requires reducing GHG emissions to 45% of their 1990 levels by 2030 and to zero by 2050, making

DOI: 10.1201/9781003404194-15

the EU the world's first carbon-neutral bloc by 2050 (Tol, 2021). According to the International Energy Agency, all countries, including those in the EU, must cease the extraction and development of new crude fossil fuels in 2021 to meet the 1.5°C climate target by 2050 (Wang et al., 2021). Furthermore, as agricultural GHG emissions, mainly methane (CH_4) and nitrous oxide (N_2O), may increase significantly in the coming decades (Hedenus et al., 2014), climate change mitigation efforts must also focus on deeply reducing these agricultural GHGs in addition to ongoing efforts to reduce GHG emissions from fossil fuel use for electricity generation, transportation, and other activities (Bryngelsson et al., 2016; Tol, 2021). Deep cuts in GHG emissions are more important than enhancing policy efforts (Tol, 2021).

Achieving sufficient agricultural production and meeting the climate target by 2050 will largely depend on Europe's ability to address several critical challenges that may hinder the successful implementation of ongoing adaptation measures and policies (Fellmann et al., 2018). Therefore, it is crucial to review figures of European agriculture under various adaptation actions and policies (Fellmann et al., 2018) to maintain productivity of croplands and animal farms, identify climate change mitigation constraints, and develop innovative actions and policies to support European farmers in meeting the challenge of climate change (Puertas et al., 2023). In this context, the main objective of this chapter is to review measures and policies that can help achieve synchrony between climate targets (i.e., achieving net-zero GHG emissions) and agricultural productivity targets (i.e., effectively increasing crop and animal production) by 2050 and identify possible challenges that may hinder their successful implementation.

12.2   EUROPEAN AGRICULTURAL ADAPTATION MEASURES

Unlike electricity generation and transportation activities, which emit large amounts of CO_2, agricultural activities are major contributors to non-CO_2 GHG emissions, particularly CH_4 and N_2O from livestock, manure management, fertilization of grasslands and croplands, rice cultivation, soil management (e.g., tillage), and burning of agricultural residues (Fellmann et al., 2018). Reducing agricultural GHG emissions is crucial for limiting global warming to 1.5°C by 2050 (Searchinger et al., 2021). For example, it has been estimated that reducing global CH4 emissions by 45% by 2030 (relative to 2010) would avoid approximately 0.3°C of warming by the 2040s (Searchinger et al., 2021). In the EU, it has been estimated that agricultural activities emitted approximately 435 million metric tons of CO_2 equivalent in 2018, accounting for about 10% of total GHG emissions (Verschuuren, 2022). Approximately 70% of these agricultural emissions were from livestock (European Commission, 2020), mostly CH4 emissions from enteric fermentation (IPCC, 2019). France, Germany, and the United Kingdom were the largest agricultural GHG emitters, accounting for approximately 45% of emissions (Fellmann et al., 2018). To meet the 1.5°C climate target by 2050 (i.e., limit the increase in global average temperature to below 1.5°C above pre-industrial levels), EU countries must make significant efforts in the coming years to reduce agricultural GHG emissions by 55% by 2030 relative to 1990 levels and achieve a balance between agricultural emissions and removals by 2050 (Verschuuren, 2022). Appropriate adaptation measures are needed to achieve these goals. These measures should enhance

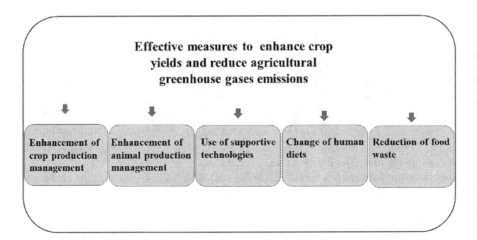

FIGURE 12.1 Effective European measures to enhance crop yields and reduce agricultural greenhouse gases emissions.

crop yields, reduce agricultural GHG emissions, and promote the capture of CO_2 from the atmosphere (Wang et al., 2021). These measures can be classified into five categories: (1) enhancement of crop production management; (2) enhancement of animal production management; (3) use of supportive technologies; (4) change of human diets; and (5) reduction of food waste (Bryngelsson et al., 2016; Wang et al., 2021) (see Figure 12.1 and Table 12.1).

TABLE 12.1

Examples of Climate Change Adaptation Measures for European Agricultural Production and Their Benefits

Adaptation Measures	Examples	Benefits
Enhancement of crop production management	Optimization of nitrogen use efficiency	Improvement of crop yield without disrupting the environment
	Efficient water use	More water savings and less soil degradation
Enhancement of animal production management	Treatment of animal farms manure	Reduction in CH_4 and N_2O emissions
	Adaptation of high-quality feed	Reduction in CH_4 and N_2O emissions
Use of supportive technologies	Precision agriculture technologies	Facilitation of agricultural practices
	Digital agriculture for N management	Identification of optimal N fertilizer application
Change of human diets	Promotion of plant-based diets	Addressing public health issues
	Less meat consumption	Reduction in GHG emissions
Reduction of food waste	Promotion of local foods	Reduction in CO_2 emissions associated with transport
	Food waste management	Less use of coal in electricity generation

12.2.1   Enhancement of Crop Production Management

Improving crop production management in croplands, for example through optimizing crop nitrogen (N) use efficiency (NUE) and efficient water use, can significantly reduce agricultural GHG emissions and increase crop yields (Shang et al., 2021; Wang et al., 2021).

12.2.1.1   Optimization of Nitrogen Use Efficiency (NUE)

The first and fundamental step toward achieving zero agricultural GHG emissions by 2050 must focus on reducing N fertilizer use in the coming years (Northrup et al., 2021). Reactive N plays a key role in supplying N to crops and soil microbes and achieving high crop yields. However, when used in excess, it can have negative environmental impacts on air and downstream water quality (Elrys et al., 2022). On average, crops and plants effectively use up to about 50% of applied N, while the rest is lost through various pathways to the surrounding environment, leading to environmental issues such as soil acidification, underground water pollution from leached nitrate (NO_3^-), and GHG emissions (Yu et al., 2015; Elrys et al., 2022). In some European agricultural areas, this percentage may reach up to 59% or higher thanks to the adoption of nutrient management plans (Winiwarter et al., 2011). Therefore, optimizing N use and developing new low-cost and environmentally friendly synthetic N fertilizer types are crucial for maximizing N benefits while reducing negative impacts on the environment (Wang et al., 2021; Elrys et al., 2022). Nitrogen use efficiency is a useful agro-environmental indicator for describing the efficiency of a crop in using N for biomass production and evaluating the environmental performance of agricultural production systems (Oenema et al., 2015). Nitrogen use efficiency values between 50% and 90% are desirable for environmental conservation and agricultural sustainability (Oenema et al., 2015; Quemada et al., 2020). Since the 1990s, the NUE of European agriculture has increased significantly due to policies such as the Nitrate Directive (91/676/CEE), but not enough to sufficiently reduce N losses to meet environmental targets in the coming decades (de Vries et al., 2021). To protect the environment from further degradation and ensure continuous agricultural production for future European populations, alternative fertilizers or reduced synthetic fertilizer input are in high demand (van Grinsven et al., 2015; Morales et al., 2022). The Farm to Fork strategy, a key component of the European Green Deal, sets ambitious goals for EU countries: compared to 2010 levels, they aim to cut nitrogen (N) losses and fertilizer use by 50% and 20%, respectively, by 2030, and even more by 2050 (de Vries et al., 2021). Usually, researchers use data on crop yield and N fertilizer to assess current NUE and also to project future changes in NUE under various N management and environmental conditions and suggest scientific recommendations and measures that can increase NUE for the agricultural production systems over the next few decades, for example, by 2050 (e.g., Ntinyari et al., 2022). At the EU level, some studies have projected future N inputs for crop production to identify local actions aimed to maintain productivity and decrease environmental issues and to suggest optimal N management measures that can help to achieve the synchrony between climate target (i.e., climate neutrality) and agricultural productivity target (i.e., increasing crop yield effectively) by 2050 (Winiwarter et al., 2011).

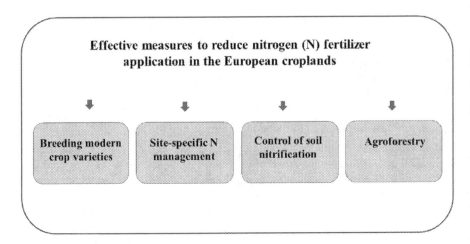

FIGURE 12.2 Effective measures to reduce nitrogen (N) fertilizer application in the European croplands.

The future effects of N on the European environment will depend on the extent of nitrogen use and NUE would increase substantially in the EU countries during the next few decades if appropriate N management measures are implemented effectively (Winiwarter et al., 2011). One way to improve crop yield limited by N without disrupting the environment is to enhance crop NUE via appropriate measures (Mălinaş et al., 2022; Govindasamy et al., 2023). The following examples are usually recommended as the appropriate measures for reducing the N fertilizer application in the European croplands (e.g., Elrys et al., 2022) (Figure 12.2).

Researchers typically utilize data on crop yield and N fertilizer to evaluate current NUE and project future changes in NUE under various N management and environmental conditions. They suggest scientific recommendations and measures to increase NUE in agricultural production systems over the next few decades, such as by 2050 (Ntinyari et al., 2022). At the EU level, some studies have projected future N inputs for crop production to identify local actions aimed at maintaining productivity while decreasing environmental issues. These studies suggest optimal N management measures to achieve synchrony between climate targets (i.e., climate neutrality) and agricultural productivity targets (i.e., effectively increasing crop yield) by 2050 (Winiwarter et al., 2011). The future impact of N on the European environment will depend on the extent of its use. If appropriate N management measures are effectively implemented, NUE could increase substantially in EU countries over the next few decades (Hutchings et al., 2020; Winiwarter et al., 2011). One approach to improving crop yield limited by N without harming the environment is to enhance crop NUE through appropriate measures (Mălinaş et al., 2022; Govindasamy et al., 2023). Examples of appropriate measures for reducing N fertilizer application in European croplands include those recommended by Elrys et al. (2022) (Figure 12.2):

- Breeding modern crop varieties with high NUE can decrease the N fertilizer application rate and reduce emissions of N_2O (Wang et al., 2021). For example, in Scotland (Western Europe), breeding modern spring barley

(*Hordeum vulgare* L.) varieties with greater NUE has increased both grain yield and N uptake (Bingham et al., 2012). Even without N fertilizer application, average grain weight was larger in modern varieties compared to older ones (Bingham et al., 2012).

- Site-specific N management is also an effective measure to increase crop N uptake, enhance NUE, and reduce N losses to the environment (e.g., Argento et al., 2021). This measure involves the correct application of N fertilizer, including the right rate, type, placement, and timing (4Rs) (e.g., Zhang et al., 2015). For instance, in Switzerland, Argento et al. (2021) showed that applying the 4Rs measure within medium to small-sized winter wheat fields enhanced NUE by about 10%, leading to a significant reduction in the amount of N fertilizer applied.

- Controlling the soil nitrification process in croplands has high potential to mitigate N losses to the environment, enhance NUE, reduce N_2O emissions, and improve crop yields (Elrys et al., 2022). This control can be achieved through the application of synthetic nitrification inhibitors (NIs) (Moir et al., 2012; Soares et al., 2023). Compared to conventional fertilizers, NIs can maintain higher soil NH_4^+ content for longer periods without negatively impacting crop yields (Akiyama et al., 2010; Soares et al., 2023). For example, controlling soil nitrification via NIs in wheat-cropping fields in northern Spain inhibited the growth of ammonia-oxidizing bacteria and reduced N_2O emissions by up to 35%, significantly improving wheat yields and environmental conservation (Huérfano et al., 2015).

- Adopting agroforestry, that is, incorporating the cultivation of woody trees into European farmlands, can play a key role in achieving zero GHG emissions in Europe by 2050 compared to conventional agriculture. Agroforestry can significantly increase carbon sequestration (up to 7.29 t C ha^{-1} yr^{-1}), decrease agricultural GHG emissions (by up to 43.4%), enhance a wide range of regulating ecosystem services, protect crops and livestock from extreme climate events, reduce soil degradation, and improve biodiversity (Kay et al., 2019). Promoting agroforestry in European farmlands over the next few decades through effective methods and policies could further decrease atmospheric GHG concentrations and mitigate environmental issues such as soil erosion and water salinization (Kay et al., 2019; Sollen-Norrlin et al., 2020; Smith et al., 2022).

12.2.1.2 Water Use Efficiency (WUE)

Water use efficiency (WUE), defined as the ratio of output (grain yield or total biomass) to the volume of water consumed by rainfed or irrigated agriculture over time, is a useful indicator for assessing the sustainability and management of regional water resources and evaluating the performance of rainfed and irrigated agriculture (Hellegers and van Halsema, 2021). Improved WUE has significantly conserved soil and water resources in European farmlands (e.g., Berbel et al., 2018). Over the past three decades, European countries, particularly those in the south, have experienced an increase in WUE due to the early implementation of relevant agricultural policies and measures. These measures include water conservation and saving technologies, soil conservation measures, adjustment of greenhouse structures, irrigation modernization (e.g., use of precise watering technology), and more efficient irrigation methods (e.g., drip irrigation), which have resulted in greater water savings and more sustainable use of soil and water resources (Berbel et al., 2018). However, as most of Europe is expected to face more frequent severe droughts in the coming decades (Lehner et al., 2017), further efforts to increase WUE are needed throughout the European agricultural domain. This is especially true for specific cropping systems such as potato systems in northern Europe and cotton and maize systems in southern Europe, where agricultural production is highly dependent on irrigation water and its effective use (Oliver et al., 2019). It is worth noting that water-saving measures may substantially reduce CH_4 emissions from European farmlands, particularly rice fields. This means that water-saving measures could be a powerful tool for mitigating CH_4 emissions from farmlands (Oliver et al., 2019).

12.2.2 ENHANCEMENT OF ANIMAL PRODUCTION MANAGEMENT

In addition to croplands, animal production farms (e.g., sheep, cattle, pork, and poultry), or livestock farms, are also a significant source of agricultural GHG emissions in European countries (Guyomard et al., 2021). In 2017, these farms contributed approximately 6.6% of total European GHG emissions (Guyomard et al., 2021), with CH_4 and N_2O, particularly problematic GHGs, being the largest emitted gases. These gases are released through enteric fermentation in ruminants (leading to CH_4 emissions) and manure management (leading to CH_4 and N_2O emissions) (Guyomard et al., 2021). Dairy and beef cattle were the largest contributors, accounting for around 80% of total livestock GHG emissions, while pork contributed 16% and poultry 4% (Guyomard et al., 2021).

Effective management of European animal production farms is needed to minimize the release of livestock GHG emissions with high global warming potential (mainly CH_4 and N_2O), slow the warming trend, and achieve net-zero carbon livestock farming by 2050. At the EU level, GHG emission reduction is currently guided by the Common Agricultural Policy, particularly through the Rural Development Program (Guyomard et al., 2021). This program has involved European livestock farmers in transitioning their production systems to minimize climatic environmental and health disservices (Guyomard et al., 2021). It has also focused on changing legal regulations and financial outlays on farms (Mielcarek-Bocheńska and Rzeźnik, 2021) and promoting effective practical measures at livestock farms. These measures include eco-innovations and climate-smart technologies, treatment and recycling of animal farm manure for plant nutrients and energy, adaptation of high-quality feed with great potential to reduce CH_4 release, and promotion of training programs to enhance farmers' managerial skills and raise climatic awareness (e.g., Stetter and Sauer, 2022). Overall, these measures have shown good performance over the past few decades and with further efforts could be promising avenues for achieving net-zero carbon livestock farming by 2050 (FAO, 2020). For example, in Germany's livestock farms, biogas production from broiler manure (via pyrolysis and biogas digestate treatment) has resulted in the lowest CH_4 and N_2O emissions, making it a good climate change mitigation strategy in the country (Kreidenweis et al., 2021). Similarly, in Switzerland,

Harter et al. (2014) found that compared to raw poultry manure, using poultry biochar can positively influence phosphorous release and reduce N_2O emissions from soils.

12.2.3 USE OF SUPPORTIVE TECHNOLOGIES

The utilization of digital tools, artificial intelligence, and precision agriculture technologies, such as digital field records, automatic steering systems, and digital irrigation systems, has significantly increased in European croplands and livestock farms over the past few decades. These technologies have been employed to address agricultural GHG emissions and enhance agricultural productivity (Gabriel and Gandorfer, 2023). In addition to their role in promoting good farming practices (Gabriel and Gandorfer, 2023), certain agricultural technologies, such as digital agriculture for nitrogen fertilizer reduction, can also play a crucial role in conserving resources and the environment, increasing soil carbon storage, and reducing European agricultural GHG emissions (Northrup et al., 2021). Bryngelsson et al. (2016) demonstrated that technologies can decrease agricultural GHG emissions by approximately 50%. Furthermore, Northrup et al. (2021) found that a combination of innovations in digital agriculture, crop and microbial genetics, and electrification could reduce GHG emissions from row crop agriculture by up to 71% (1,744 kg CO_2 eq. ha^{-1}) within the next 15 years. Given these promising results, it would be beneficial to enhance the ability of European farmers to effectively implement agricultural technologies through training sessions and on-farm demonstrations (Chen and Chen, 2022). Such training is expected to facilitate the adoption of complex and challenging agricultural technologies by farmers (Chen and Chen, 2022).

12.2.4 CHANGE OF HUMAN DIETS

In the European countries, the livestock sector accounts for approximately 6.6% of total European GHG emissions (Guyomard et al., 2021) and is a significant source of CH_4 and N_2O, particularly problematic GHGs (Gerber et al., 2013). Given that vegetable protein sources produce lower GHG emissions than animal protein sources, especially beef and mutton (Davis et al., 2010), a substantial shift in current dietary habits toward reduced meat consumption and increased plant-based diets could be beneficial in reducing CH_4 and other GHG emissions, achieving climate targets by 2050, addressing public health issues, and promoting animal welfare (Kwasny et al., 2022). However, there are few studies on the potential to reduce GHG emissions through changes in human diets (Bryngelsson et al., 2016). For example, Green et al. (2015) found that promoting diets among UK adults that conform to World Health Organization (WHO) guidelines could reduce GHG emissions from food by 17% by 2050 (relative to 1990). It was anticipated that CH_4 and N_2O emissions from food consumption could be reduced to the extent necessary to meet the EU climate target for 2050. However, promoting healthier diets among consumers in the coming years may not be an easy task as meat and other food consumption is often rooted in cultural practices, societal norms, and daily habits (Stoll-Kleemann and Schmidt, 2017). This suggests that urgent and stronger actions and policies are needed to raise awareness regarding food consumption and its impact on climate change.

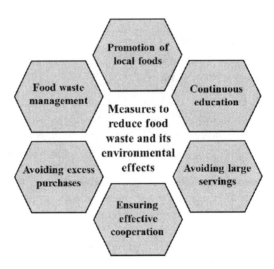

FIGURE 12.3   Useful measures to reduce food waste and its associated environmental effects.

12.2.5 REDUCTION OF FOOD WASTE

Approximately 88 million tons of food are wasted annually in the EU, representing up to 16% of the total GHG emissions from consumed food (Scherhaufer et al., 2018; Jeswani et al., 2021). Reducing food waste, one of the Sustainable Development Goals (SDGs), can therefore help to decrease GHG emissions (von Braun et al., 2023). There are several effective options for reducing food waste at the local level (Figure 12.3), including:

- Promotion of local foods: At the local level, food is often wasted due to damage or spoilage caused by long supply chains and transportation processes that consume more external resources (Yang et al., 2023). Promoting eating habits that favor locally produced foods can reduce CO_2 emissions associated with transport activities, decrease transportation costs, and minimize waste caused by food spoilage during transport (Yang et al., 2023).
- Food waste management: Organic food waste can be biologically fermented to produce methane, which can then be burned to generate electricity. This reduces the use of unsustainable sources such as coal for electricity generation and its associated CO_2 emissions (Kaur et al., 2019; Yang et al., 2023).
- Continuous education: Educating consumers across all age groups, particularly young people, on how small actions such as proper management of food waste can make a significant difference in mitigating GHG emissions (von Braun et al., 2023).
- Avoiding excess purchases: Reducing purchases of excessive quantities of food, especially those with higher environmental impacts than vegetables (e.g., meat and cheese), is an effective measure to reduce GHG emissions from food consumption and related waste (Jeswani et al., 2021).

- Avoiding large servings in the hospitality sector: Reducing serving sizes in the hospitality sector can help to decrease food waste and its associated environmental impacts (Jeswani et al., 2021).
- Ensuring effective cooperation: Engaging all actors in the food supply chain, including food industries, retailers, and consumers, can help to reduce waste generation and ensure effective management of food waste (Jeswani et al., 2021).

12.3 CONSTRAINTS, LIMITATIONS, AND RECOMMENDATIONS

Currently, all EU countries are increasingly implementing various agricultural adaptation measures and policies, such as the Common Agricultural Policy, Precision Agriculture, European Green Deal, agroforestry, and soil conservation measures, to provide environmental goods and services beyond food production (Candel and Biesbroek, 2018). These measures and policies have been found to be effective in increasing agricultural production and mitigating GHG emissions (Candel and Biesbroek, 2018). However, there are still many research gaps, constraints, and limitations that need to be addressed in the coming years to help EU decision-makers achieve synchrony between climate targets (i.e., achieving net-zero GHG emissions) and agricultural productivity targets (i.e., effectively increasing crop and animal production) by 2050. Some examples include:

- Agricultural technologies have been extensively used and evaluated in large-scale crop production and livestock farming systems in EU countries. However, the efficacy of using digital technologies in small-scale farming has rarely been investigated (e.g., Gabriel and Gandorfer, 2023). Further research on this topic is recommended.
- Although meat consumption can contribute to various health and environmental issues (e.g., increased GHG emissions), most EU countries have shown little interest in making major agricultural policy changes to reduce meat production and encourage consumers toward environmentally friendly food choices (e.g., Vittersø et al., 2015; Austgulen et al., 2018). This suggests that urgent and stronger actions and policies are needed in this area. Providing consumers with WHO dietary recommendations along with educational courses on environmentally friendly food choices is recommended (Kwasny et al., 2022). Food suppliers and supermarkets can also play a crucial role by increasing the attractiveness of plant-based diets (Yang et al., 2023).
- Reducing food waste can also play a role in reducing GHG emissions. However, accurate information on food waste at the EU level (e.g., sources, causes, amounts) is still scarce (Jeswani et al., 2021), which may make the development and implementation of effective waste prevention and management measures difficult (Jeswani et al., 2021). Therefore, collecting sufficient data on European food waste is necessary to implement appropriate waste prevention measures and monitor their progress (Jeswani et al., 2021). Food waste prevention measures and targets should not only be developed

based on waste quantities but also on their contribution to reducing GHG emissions (Jeswani et al., 2021).

- Although agroforestry is a sustainable land use practice that increases carbon storage in soils, reduces GHG emissions, and enhances ecosystem service delivery from croplands compared to treeless croplands (Mosquera-Losada et al., 2023), the adoption of this practice across some EU countries still faces technical (e.g., lack of knowledge on the best combinations adapted to specific field levels), economic (e.g., more research on the economic benefits of agroforestry is needed), educational (e.g., farmers are not aware of the role of agroforestry in their agriculture), and policy development challenges (e.g., maintenance of current agroforestry systems) (Mosquera-Losada et al., 2023). Addressing these challenges would be beneficial in achieving climate neutrality in Europe by 2050. The policy recommendations suggested by Mosquera-Losada et al. (2023) could be helpful in overcoming these challenges.

12.4 CONCLUSIONS

At the EU level, agriculture in the coming decades must integrate practices with low environmental impact (i.e., reduced GHG emissions) to sustainably feed the growing population without contributing to further warming. This requires a transition to eco-intensification of agriculture. To facilitate this transition, all EU countries are currently implementing various agricultural adaptation measures and policies, such as the Common Agricultural Policy, Precision Agriculture, European Green Deal, agroforestry, and soil conservation measures. Although the majority of these adaptation measures and policies are performing relatively well, there is a strong need for regular assessment of their performance to identify any limitations and constraints that may hinder the achievement of synchrony between climate targets (i.e., achieving net-zero GHG emissions) and agricultural productivity targets (i.e., effectively increasing crop and animal production) by 2050. Gathering performance information from different research disciplines and pedoclimatic conditions across EU countries would be beneficial in enhancing current agricultural adaptation strategies and monitoring their progress.

13 Effective Agricultural Adaptation Measures to Support Africa in Achieving Food Security under Climate Change

13.1 INTRODUCTION

Climate change is emerging as a major environmental issue for the African continent, resulting in reduced crop and animal production, increased food insecurity, hunger and poverty, and deterioration of soil resources through degradation processes such as erosion and salinity (Akinnagbe and Irohibe, 2014; Salack et al., 2022). Without adequate intervention in the coming years, the projected changes in climate by 2050 will further exacerbate these effects (Emediegwu et al., 2022). For example, relative to current data, it is estimated that African crop production will decrease by 13%, 11%, and 8% in West and Central Africa, North Africa, and East and Southern Africa respectively by 2050 (WHO, 2020). Furthermore, some major crops such as wheat may experience a critical reduction of up to 21% (WHO, 2020). Moreover, according to a recent report by the UN's International Fund for Agricultural Development (IFAD), if no changes are made to African agricultural practices, yields of staple crops in some areas of East and Southern Africa could decrease by up to 80% by 2050 (IFAD, 2021). Therefore, the operational implementation of effective adaptation measures is crucial for enabling farmers and decision makers to respond to the effects associated with climate change and exploit its transformative opportunities (Salack et al., 2022). Currently, several agricultural adaptation measures have been implemented by local African communities to reduce the impact of climate change on crop and animal production. However, most have been applied in the context of small-scale farming (Fonta et al., 2018; Salack et al., 2022). Climate change effects on crop production have been addressed through planting climate-tolerant crops, optimizing growing conditions (e.g., changing planting times), adopting soil water conservation practices, improving irrigation water management, promoting organic agriculture, practicing agroforestry, and harvesting rainwater (Akinnagbe and Irohibe, 2014; Rhodes et al., 2014; Muchuru and Nhamo, 2019). Climate change effects on animal production have been mitigated through changes in practices such as diversification, intensification, and agro-pastoralism (Akinnagbe and Irohibe, 2014; Salack et al., 2022).

DOI: 10.1201/9781003404194-16

Despite increasing global efforts to fund and implement adaptation measures and their positive effects on African agriculture, challenges and barriers such as inadequate agricultural infrastructure, insufficient funding for adaptation, weak implementation of adaptation measures in large-scale farming, lack of skills to formulate bankable and result-oriented climate actions, and low use of adaptive technologies still exist. These may hinder the successful implementation of effective adaptation options in the coming years (Akinnagbe and Irohibe, 2014; Emediegwu et al., 2022; Salack et al., 2022). As Africa is expected to face critical changes in climate by 2050, such as increased warming trends, frequent droughts, and rainfall variability, it is essential for African agriculture to become more adaptive to these potential changes through the implementation of effective adaptation measures in the coming years (Girvetz et al., 2019). Otherwise, climate change in 2050 coupled with a significant population and food demand growth may result in many African countries experiencing insufficient agricultural productivity and severe malnutrition and food insecurity issues (Beltran-Peña and D'Odorico, 2022). Regular review of available adaptation measures in Africa is necessary to identify potential challenges, develop the most effective adaptation measures and monitor their progress, and support policymakers in making decisive steps toward climate change adaptation across the entire African continent (Akinnagbe and Irohibe, 2014). In this context, the main objective of this chapter is to review current African agricultural adaptation measures and identify the most effective ones that can help reduce the effects of climate change on African agricultural productivity and promote sustainable agriculture by 2050.

13.2   AFRICAN AGRICULTURAL ADAPTATION MEASURES

To effectively respond to the long-term effects of climate change, agricultural management practices must be transformed into more sustainable and effective ones (IFAD, 2021). A successful transition is expected to improve crop production, build resilience in farming systems, and reduce harm to the environment (Nyasimi et al., 2014). Numerous effective measures, such as crop diversification, improvement of irrigation efficiency, organic agriculture, planting of climate-tolerant crops, adoption of soil water conservation practices, agroforestry, rainwater harvesting, and sustainable livestock management, are being promoted in Africa to reduce climate change-related risks while improving crop and animal production (Akinnagbe and Irohibe, 2014; IFAD, 2021) (Figure 13.1). This section analyzes these effective measures individually, focusing on their strengths, potential constraints that may hinder their successful implementation, and recommendations for improved performance.

13.2.1   CROP DIVERSIFICATION

Crop diversification refers to the practice of planting more than one variety of crops in a given agricultural area through various cropping activities such as crop rotations and intercropping (Makate et al., 2016; Beillouin et al., 2021). It is an ecologically feasible and cost-effective adaptation measure (Makate et al., 2016) that is highly necessary for agricultural areas expected to experience severe climate change in the coming years (Rippke et al., 2016). Compared to monoculture, crop diversification

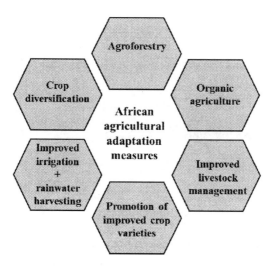

FIGURE 13.1 Climate change adaptation measures for African agricultural production.

has a much greater potential to reduce risks associated with agricultural production, improve soil fertility, reduce pests and diseases, enrich on-farm biodiversity, and increase crop yield per unit area (Joshi, 2005; Lin 2011; Makate et al., 2016; Beillouin et al., 2021). Adopting crop diversification practices under current climate conditions has resulted in numerous benefits in many African countries such as Tanzania, Kenya, Zimbabwe, Rwanda, Ethiopia, Niger, Zambia, and Ghana. For example, in Zimbabwe, Makate et al. (2016) found that wider adoption of diversified cropping systems significantly increased legume and maize productivity, farmer income, food security, and household nutrition. Similarly, in northern Ghana, recent studies (e.g., Danso-Abbeam et al., 2021; Baba and Abdulai, 2021) have demonstrated significant positive correlations between crop diversification and household food security. However, it is important to note that the potential for transitioning to diversified cropping systems depends on the ability of the selected crops to be sustained under the climate of the given areas where they are grown (Chemura et al., 2022). Overall, Waha et al. (2018) found that on a continental scale, the greatest potential for crop diversification is located in agricultural areas with annual rainfall of 500 to 1,000 mm and rainfall variability of 17% to 22%.

13.2.2 RAINWATER HARVESTING (RWH)

As most African countries are expected to experience critical changes in rainfall patterns during crop growing seasons (e.g., rainfall shortages in some areas and intense rainfall events in others) due to climate change in the coming years (Lebel et al., 2015), the development of small-scale irrigation systems and rainwater harvesting (RWH) structures (Figure 13.2) could be a promising measure for enhancing precipitation capture, increasing water availability, reducing irrigation costs, and improving water management within small-scale farming, particularly in areas that

FIGURE 13.2   A schematic diagram for a typical rainwater harvesting system.

receive insufficient rainfall to grow crops (Rockström and Falkenmark, 2015). The most common small-scale water harvesting methods in Africa include terracing to conserve soil moisture (as in Kenya), using dams and ditches to channel runoff into fields (as in Eritrea), and using on-farm storage facilities such as ponds, tanks, and sub-surface storage in soil (as in South Africa, Ethiopia, and Sudan) (Rockström and Falkenmark, 2015). Rainwater harvesting has significantly increased yields of major food crops in many regions of Africa. For example, in Northern Togo, it has demonstrated great potential for maize production during the dry season (Gadédjisso-Tossou et al., 2020). Overall, an assessment study at the African continental scale conducted by Lebel et al. (2015) found that RWH could bridge approximately 40% and 31% of maize yield gaps attributable to water deficits under current and future (2050) climates respectively, thereby reducing future irrigation water requirements. The same study also found that by bridging water deficits across Africa, RWH could improve maize yields by up to 50% by 2050 (relative to the baseline period of 1986–2005). As maize is the most widely grown crop in Africa (Portmann et al., 2010) and most smallholder farmers are poor and unable to use more advanced agricultural systems to conserve rainwater (e.g., Khapayi and Celliers, 2016), RWH could be a promising measure for increasing crop yields with relatively low costs and enhancing future food security for African people (Leal Filho et al., 2022). However, it is important to note that some key technical (e.g., deterioration of rainwater quality) and economic (e.g., financial issues) barriers may hinder the expansion of RWH implementation in Africa. Fortunately, many solutions are available to overcome these barriers. For instance, deterioration of rainwater quality due to unsuitable types and conditions of collection, delivery, and storage can be avoided through simple technical solutions such as the use of screens and first flush tanks (e.g., Coombes, 2015). Improving individual and community financial capacity could also help overcome economic barriers.

13.2.3   IMPROVED IRRIGATION

In Africa, agricultural productivity is highly dependent on rainfall (Emediegwu et al., 2022), and only 9.4% of the total cropland (24.5 Mha) was under irrigation in 2014 (Xiong et al., 2017). However, the irrigation potential in this continent (from

TABLE 13.1

Examples of Efficient Irrigation Practices to Improved Agricultural Productivity in Africa

Measure/Practice	Reference
To apply water in a reasonable amount, at the right time, and uniformly across the entire irrigated area in the most economical way possible	Reinders (2011)
To Implement solar-based irrigation systems in dry lands	Birhanu et al. (2023)
To apply drip irrigation via proper irrigation schedule in arid agricultural areas	Ouédraogo et al. (2021)
To promote innovative methods in drylands that can help to conserve rainwater in situ or harvest it for use elsewhere	Stroosnijder et al. (2012)
To avoid soil salinity risk after irrigation through practicing efficient drainage and growing salt-resistant crops (e.g., halophytes) together with conventional crops	Mohanavelu et al. (2021)
To Promote innovative water-use technologies for improving water accounting and optimizing food production without increasing water applied	Nhamo et al. (2023)

renewable groundwater) could reach up to approximately 48% of cropland (Altchenko and Villholth, 2015), which could significantly improve food security and help manage risks associated with increasingly unreliable rainfall due to climate change (Uhlenbrook et al., 2022). Therefore, expanding irrigation onto both new croplands and currently underperforming rainfed croplands, along with improved irrigation management within African smallholder irrigated schemes through effective changes in irrigation practices (Table 13.1), could be a powerful tool to mitigate climate change (AGRA, 2019), boost agricultural productivity by approximately 50% (AGRA, 2019), and overcome extreme poverty (Tramberend et al., 2021). Due to the great potential to scale up irrigation, recent years have seen continued investments in new irrigation schemes across many African countries, leading to significant increases in crop yields and farmer income (AGRA, 2019). For example, in Malawi, expanding the area of arable land under irrigation has increased farmer incomes by 65% (AGRA, 2019). However, to ensure high performance of new African irrigated schemes, factors such as inadequate maintenance of irrigation infrastructure, unsuitable design of irrigation schemes, lack of technical support, weak motivation of local people, and lack of a regulatory framework that clarifies the roles of different stakeholders (e.g., government, water user associations, and farmers) must be avoided (Nalumu et al., 2021). Moreover, due to projected changes in climate in Africa by 2050 (e.g., more extreme climate events and increased intra-annual rainfall variability) (Almazroui et al., 2020a), it is essential to enhance the stability and performance of irrigated schemes through the promotion of on-farm rainwater harvesting practices, increased investment in soil-water conservation measures, shifting to more efficient irrigation methods such as drip irrigation, installing photovoltaic panels over croplands to mitigate evaporation, promoting efficient drainage systems to avoid soil salinity issues, and planting less water-intensive crops (Rosa, 2022) (Figure 13.3). Additionally, practices such as irrigation scheduling, deficit irrigation, optimal application of fertilizers, and conservation tillage can significantly improve irrigation in Africa by reducing water

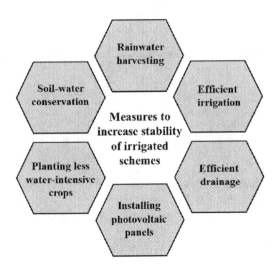

FIGURE 13.3 Measures to increase stability of irrigated schemes in Africa.

losses within irrigated croplands and increasing crop yields simultaneously (Shock et al., 2013). For instance, in Egypt, Ibrahim et al. (2016) found that applying a deficit irrigation amount equal to 80% of maize evapotranspiration via drip method together with optimal fertilizer application at the end of the irrigation event can reduce irrigation water losses by 27% and provide higher maize yield compared to conventional irrigation (full irrigation).

13.2.4 AGROFORESTRY

Agroforestry, the practice of combining agriculture and forestry, is a powerful option for avoiding deforestation, enhancing carbon sequestration and, reducing GHG emissions to mitigate climate change (Kim et al., 2016; Ma et al., 2020; Mayer et al., 2022). Compared to conventional agriculture, the adoption of agroforestry in Africa over the past few decades has resulted in numerous agricultural and environmental benefits such as climate change mitigation, biodiversity conservation, significant increases in crop yield, notable enhancements of soil properties (especially soil organic carbon, soil water content, and infiltration), reductions in land degradation and runoff and soil loss (Mbow et al., 2014; Kiyani et al., 2017; Kuyah et al., 2019). Furthermore, agroforestry has improved farmer income by providing additional products for sale or home consumption (Thangataa and Hildebrand, 2012; Mukhlis et al., 2022). Kuyah et al. (2019) analyzed the effects of agroforestry on crop yields and soil organic carbon across sub-Saharan African countries and found that compared to conventional agriculture, agroforestry increased crop yield and soil organic carbon by 54–85% and 58%–68% respectively, depending on the country. The same study showed that in most of the studied areas, yield increased sufficiently to offset decreases caused by the presence of trees within croplands. However, despite these promising findings, the adoption of agroforestry has not been widespread in many regions of Africa due to factors related to the performance of agroforestry practices, political and socioeconomic environment, lack of skills and technical support, and weak participation by many farmers in planting trees on their farms (Mbow et al., 2014; Kim et al., 2021). Therefore, there is an urgent need to address these factors in the coming years to achieve sustainable agriculture and meet climate neutrality targets by 2050. African governments must make greater efforts toward more effective adoption of agroforestry (Kiyani et al., 2017). This is crucial as 71% of African countries have committed to adopting agroforestry for climate change mitigation and more than 50% of countries on the African continent (28 out of 54) have committed to ecologically restoring approximately 1,130,000 km^2 of land by 2030, mainly through agroforestry (Rosenstock et al., 2019).

13.2.5 PROMOTION OF IMPROVED CROP VARIETIES

In Africa, the cultivation of traditional crop varieties that are not resilient to the emerging threat of climate change has led to insufficient food production, malnutrition, and low household incomes (Okori et al., 2022). As a result, high-yielding and climate-tolerant crop varieties that can produce good yields even under harsh conditions have recently gained popularity among farmers across Africa (Okori et al., 2022). For example, in Tanzania, where maize production must increase by 2.4% annually to meet food demand (Msangi and You, 2010), new maize varieties (CZH132019Q and CZH132003Q) have been approved for their high tolerance to drought events and strong resistance to common maize diseases (e.g., gray leaf spot and streak virus disease), resulting in higher grain yields (up to 31%) compared to traditional varieties (Okori et al., 2022). Therefore, as maize is the most important staple food in Tanzania, promoting these new varieties in the coming years will significantly enhance food security (Okori et al., 2022). Similarly, in Malawi, two new drought-tolerant bean varieties (SER83 and SEN43) have shown higher grain yields (up to 48%) under rainfed conditions compared to an old traditional bean variety (Napilira) (Okori et al., 2022). Furthermore, the strong resistance of these new bean varieties to diseases makes them popular with farmers as it reduces the need for costly pesticide applications (Okori et al., 2022). In Kenya, where potato is the second most important staple food (Muthoni et al., 2017) and can yield up to 50 t ha^{-1}, promoting climate-resilient potato varieties in the coming years has great potential to increase farmers' resilience to climate change and significantly improve potato production (Kimathi et al., 2021). Overall, it is important to note that some agronomic practices such as crop rotation, minimum tillage, and soil conservation measures are necessary to maximize the benefits of adopting high-yielding and climate-tolerant crop varieties (Kimathi et al., 2021).

13.2.6 ORGANIC AGRICULTURE

As the intensive use of synthetic fertilizers (e.g., phosphates) in agriculture is a major contributor to nutrient pollution, there is no doubt that a transition toward organo-mineral fertilization or organic agriculture can be a good alternative for protecting soil and water resources from nutrient pollution and achieving synchrony between climate change mitigation and enhancement of crop production (Neuhoff and Kwesiga, 2021).

Organic agriculture is an eco-friendly agricultural system that excludes the use of synthetic fertilizers for crop production and instead uses biological fertilizers derived from animal or plant waste (e.g., animal and green manure) (Migliorini and Wezel, 2017), contributing to higher soil organic carbon content and lower GHG emissions from agricultural activities compared to nonorganic systems (Holka et al., 2022). Moreover, when combined with other environmentally friendly farming practices, notable reductions in GHG emissions can be achieved (Holka et al., 2022). In addition to its key role in reducing GHG emissions (through organic carbon sequestration), the practice of organic agriculture in some African areas over the past few years has had positive effects on crop yields (Neuhoff and Kwesiga, 2021). This can be attributed to the fact that higher amounts of organic matter in soil lead to greater water-holding capacity (i.e., soil retains more water), higher soil nitrogen retention, and increased microbial activity, resulting in better crop yields, especially during extreme climate events such as droughts and heatwaves (Hoorfar, 2014). This demonstrates the great potential of organic agriculture to increase crop yields in a future climate. For example, Neuhoff and Kwesiga (2021) analyzed paddy maize productivity from 2015 to 2018 in Uganda and found that optimal application of poultry and green manure (approximately 120 kg N ha^{-1}) increased maize yield from 1.6 t ha^{-1} (in non-organic systems) to 5 t ha^{-1}. However, to maximize its benefits, organic agriculture must be supported by the implementation of good agricultural practices such as minimum tillage and soil conservation measures (Figure 13.4) (Kwesiga et al., 2019). In the same context, in pomegranate fields in Tunisia, Haj-Amor et al. (2020) showed that applying an optimal amount of cow manure (3,748 kg ha^{-1} year^{-1}) significantly enhanced soil characteristics (particularly moisture, organic carbon content, and macronutrients), increasing resistance to future climate change by 2050. Globally, in 2020, organic agriculture was practiced on approximately 75 million ha of agricultural land, representing about 1.5% of total agricultural land (Holka et al., 2022). This percentage was much lower in Africa where only 0.2% of agricultural land is dedicated to organic farming (De Bon et al., 2018).

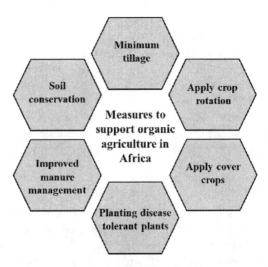

FIGURE 13.4 Necessary measures to support organic agriculture in Africa.

To increase this percentage, constraints such as scarcity of animal manure, unsuitable soil properties in some areas, limited knowledge among farmers, and limited funding opportunities must be addressed as soon as possible through continental, national, and regional policies (e.g., De Bon et al., 2018; Neuhoff and Kwesiga, 2021). These policies must involve all actors concerned with organic agriculture and be supported by international development organizations due to the weak financial capacity of some African countries (De Bon et al., 2018).

13.2.7 IMPROVED LIVESTOCK MANAGEMENT

According to Tubiello et al. (2014), approximately 70% of African agricultural GHG emissions come from the livestock sector. Despite contributing only 4% to global milk production from dairy cattle, unsuitable management practices such as overgrazing, low feed efficiency, and burning of deteriorated forage and biomass residues result in the African livestock sector emitting 10% of global enteric CH_4 (Balehegn et al., 2021). With a growing population and increasing demand for milk in most African countries by 2050 (Alexandratos and Bruinsma, 2012), continued unsuitable management practices may lead to further CH_4 emissions (Balehegn et al., 2021). Therefore, effective measures to reduce CH_4 and other GHG emissions from the African livestock sector while improving productivity are urgently needed. Adaptation measures such as increasing feed efficiency through the use of forages with low CH_4 emissions, range and grazing land restoration, proper manure management, and enhancement of herd genetics can significantly decrease GHG emissions while improving productivity (Balehegn et al., 2021). For example, in Ethiopia, Berhanu et al. (2019) found that three tropical forages – *M. stenopetala*, *C. juncea*, and *L. leucocephala* – can decrease CH_4 emissions by up to 25% while providing valuable feed resources for ruminants. In East Africa, improving carbon sequestration in rangelands through moderate grazing, restoration of degraded grasslands, and establishment of pasture enclosures has the potential to mitigate climate change and enhance livestock productivity (Mgalula et al., 2021). Specifically, grassland regeneration and restoration of degraded lands in Rwanda, Tanzania, Uganda, Ethiopia, Burundi, and Kenya have contributed to soil organic carbon stocks ranging from 0.1 to 93 Mg C ha^{-1} (Tessema et al., 2019; Mgalula et al., 2021). However, due to a scarcity of data on GHG emissions from livestock production systems in Africa and inconsistencies in available information, further research and accurate data are urgently needed across all African countries to guide mitigation measures (Mgalula et al., 2021).

13.3 CONCLUSIONS

This chapter concludes that Africa may face critical effects of climate change on crop and animal production by 2050. Therefore, it is highly and urgently necessary to improve and strengthen the agriculture sector through the implementation of successful adaptation and mitigation measures, provision of financial and technical support, education, and outreach programs to avoid the worst food security scenarios. Adaptation measures should be implemented in close association with farmers to ensure high performance. However, several factors such as weak knowledge among farmers and scarcity of data may negatively influence the adoption of specific adaptation strategies by smallholder farmers. Additionally, political and economic constraints such as unsuitable agricultural infrastructure, insufficient funding for adaptation, and lack of legal frameworks may hinder the successful implementation of adaptation measures. Therefore, greater effort from African governments and international organizations is required to strengthen the provision of agricultural extension services, provide training on best agronomic practices under climate change for farmers, and ensure sufficient funding.

14 Enhancing Agricultural Production and Reducing Greenhouse Gas Emissions in Asia through Adaptation Measures

14.1 INTRODUCTION

In addition to its significant contribution to economic growth, agriculture in Asia plays a crucial role in society by reducing hunger, increasing the income of farmers (especially smallholders), and enhancing food security (Nor Diana et al., 2022). However, over the past few decades, the effects of climate change, including rising sea levels, flooding, extreme weather events (i.e., droughts and heatwaves), and rainfall variability have negatively impacted agricultural performance. This has resulted in substantial crop yield losses, decreased farming incomes, and loss of employment opportunities in many regions of Asia (Nor Diana et al., 2022). For example, in China, the world's leading producer of rice (*Oryza sativa* L.), rice yields decreased by approximately 12% between 1961 and 2010 due to climate change (Saud et al., 2022). Similarly, in South Asia, the effects of climate change on fundamental cereal crops have generally been negative and significant (Knox et al., 2012). In India, the second-largest producer of wheat (*Triticum aestivum* L.), current climate conditions are outside the optimal range for wheat growth, resulting in substantial yield losses due to the crop's high vulnerability to climate change (Daloz et al., 2021). Gupta et al. (2017) found that despite adaptation efforts in India, climate change reduced wheat yields by 5.2% between 1981 and 2009. Ortiz et al. (2008) predicted that climate change may render a large area of India's Indo-Gangetic Plains, a major wheat-growing region, unsuitable for wheat cultivation by 2050. In Southeast Asia, climate change has reduced total agricultural employment by up to 20% between 2014 and 2016 (The World Bank, 2022).

Climate change in Asia has continued to worsen, and it is imperative to take adequate action to preserve crop and animal production and ensure food security (Nor Diana et al., 2022). Without appropriate intervention in the near future, previous studies (e.g., Knox et al., 2012; Li et al., 2022) predict that the effects of climate change on Asian agriculture by 2050 will be even more severe than they are currently. Effective intervention is urgently needed in Asia due to factors such as the large dependence of agricultural production on rainfall and its high vulnerability to even small changes in climate, the crucial role of agriculture in supporting the livelihoods of Asian people, the limited capacity of farmers to rapidly adapt (due to small farm sizes), the rapidly growing population and its associated impacts such as increased food demand and pressure on soil and water resources, and the lack of suitable institutions and policies to mitigate climate risks in agriculture in many regions (Aryal et al., 2020) and safeguard agricultural production (Amare and Simane, 2018). Measures to adapt to climate change by 2050 must be implemented as soon as possible and aligned with current and future climate conditions (Alam et al., 2013).

Many efforts and projects have been initiated to develop and test adaptation measures to enhance Asian agricultural production and help farmers adapt to changing climatic conditions in the coming decades (Nor Diana et al., 2022). However, there is a need to strengthen these measures through increased institutional and technical support (Aryal et al., 2020). Additionally, given that current agricultural production practices in Asia are typically greenhouse gas (GHG) intensive, there is a need to substantially reduce GHG emissions from Asian agriculture (Aryal, 2022). As such, a reconsideration of Asian climate policies is necessary to achieve a balance between GHG emission reduction and sufficient agricultural production by 2050 (Aryal et al., 2020; Aryal, 2022). A detailed review of current agricultural adaptation measures in Asia would be useful in assessing their performance, identifying potential constraints and factors that may hinder their successful implementation, and making recommendations to promote adaptation progress and design better climate policies for 2050 (Aryal et al., 2021). In this context, the main objective of this chapter is to conduct such a detailed review.

14.2   ASIAN AGRICULTURAL ADAPTATION MEASURES

Rapid population growth and climate change in Asia are expected to increase food demand by 40% to 50% by 2050 (e.g., Bodirsky et al., 2015). Meeting this future demand will be a significant challenge, as Asian countries currently contribute over 50% of the world's total GHG emissions (Azhgaliyeva and Rahut, 2022), a large proportion of suitable land for farming is already in use (e.g., in South Asia, 94% of land is under production), and many agricultural areas are facing issues such as water shortages, land degradation, and extreme weather events (Amarnath et al., 2017; Jat et al., 2022). However, by identifying and implementing effective adaptation measures, it is possible to mitigate some of the negative effects of climate change on Asian agricultural production and increase both crop and animal production while reducing GHG emissions from the agriculture sector (Huppmann et al., 2018). Numerous effective measures, such as improved rice field management, optimized crop rotation, enhanced crop water productivity (CWP), agroforestry, biofuel production, and sustainable livestock management are being promoted in Asia to reduce climate-related risks while improving crop and animal production (e.g., Aryal et al., 2020) (Figure 14.1). This section analyzes these measures individually, discussing their strengths, potential constraints that may hinder their successful implementation, and recommendations for improved performance.

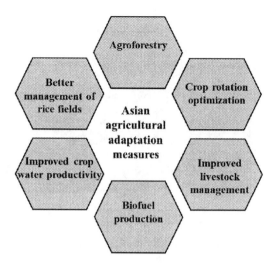

FIGURE 14.1 Climate change adaptation measures for Asian agricultural production.

14.2.1 BETTER MANAGEMENT OF RICE FIELDS

In addition to its crucial role in ensuring food security for Asian countries, rice also plays a fundamental role in enhancing global food security. However, the significant methane (CH_4) emissions associated with rice cultivation are a major drawback. CH_4 is one of the most potent GHGs and may account for up to 30% of global agricultural GHG emissions (Gupta et al., 2021). Rice cultivation is the second-largest contributor to CH_4 emissions among agricultural activities, after enteric fermentation (Smartt et al., 2016). Rice is typically grown in continuously flooded soil conditions, which cut off the oxygen supply from the atmosphere to the soil and result in anaerobic fermentation. This process leads to the proliferation of methanogenic bacteria (methanogens) that produce CH_4 and release it into the atmosphere (Smartt et al., 2016). Due to its extensive rice cultivation, Asia is one of the largest CH_4 emitters in the world (Ouyang et al., 2023). Ouyang et al. (2023) found that rice CH_4 emissions across Monsoon Asia were approximately 20 Tg per year between 2001 and 2015. Given this significant emission and its potential contribution to global warming, there is a need to reduce CH_4 emissions from Asian rice fields through appropriate measures. Currently, there are several measures that can both increase rice production (rice yields must increase by 28% to meet global demand by 2050; Alexandratos and Bruinsma, 2012) and reduce CH_4 emissions from Asian rice fields. The most effective measures include transitioning to new rice varieties, optimizing fertilization, implementing water-saving practices, and improving tillage practices (Theint et al., 2015; Fan et al., 2020; Li et al., 2022) (Figure 14.2):

- Transition to new rice varieties
  Over the past two decades, numerous studies have assessed long-term rice yield gains in various Asian countries (Kumar et al., 2021; Li et al., 2022; Rahman et al., 2023). These studies have found that transitioning to new

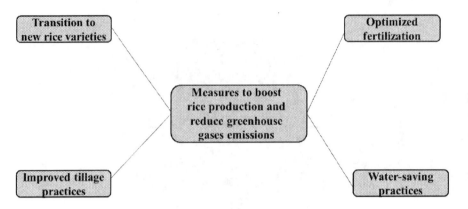

FIGURE 14.2 Measures to increase Asian rice production and reduce greenhouse gas emissions.

rice varieties can result in increased yields and reduced CH_4 emissions compared to older varieties. For example, in the Yangtze River Basin, China's largest rice-producing region, grain yields of 2010s cultivars (*Wuyunjing 27* and *Nanjing 9108*) increased by up to 93.9% compared to 1950s rice varieties (*Huangkezao* and *Guihuaqiu*), while CH_4 emissions decreased by up to 41.2% (Li et al., 2022). These positive outcomes were primarily attributed to improvements in root traits and optimization of photosynthate allocation to biomass and grain yields (Li et al., 2022). In South Asia, Mottaleb et al. (2017) found that adopting drought-tolerant rice varieties can increase rice yields by up to 9%. In coastal agricultural areas of Bangladesh, where soil salinity is a major land degradation issue, promoting saline-tolerant rice varieties (such as BRRI Dhan 11, BRRI Dhan 28, BRRI Dhan 29, and CSR 26) is a promising option for enhancing rice yields under climate change and its associated effects on soil salinity (Aryal et al., 2020).

• Water-saving practices

The intensive use of water in rice fields, such as continuous flooding (CF), is a major contributor to GHG emissions (Smartt et al., 2016). As such, water-saving practices such as alternative wetting and drying (AWD), dry cultivation (D), and intermittent irrigation can provide an opportunity to reduce GHG emissions (Carrijo et al., 2017; Akter et al., 2018; Faiz-ul Islam et al., 2020). For example, in Bangladesh, Akter et al. (2018) found that AWD can reduce GHG emissions ($CO_2 + CH_4$) by 50% compared to CF, leading to less soil organic matter degradation. However, some studies (e.g., Carrijo et al., 2017; Zhang et al., 2021) have suggested that these water-saving practices can decrease rice yields due to the low drought tolerance of rice varieties. As an alternative option, Zhang et al. (2021) proposed that combining water-saving practices with drought-resistant rice (WDR) could significantly reduce GHG emissions while maintaining rice yields. Specifically, they found that compared to a common rice variety, the WDR variety increased rice yields by 24% under D practice and that AWD practice reduced CH_4 emissions by

7 to 64% compared to CF practice. This suggests that the D-WDR mode can significantly reduce GHG emissions without compromising rice yields, making it a promising climate change mitigation option (Zhang et al., 2021).

- Optimized fertilization methods

  Among all fertilizers, nitrogen (N) is the most essential for crop production (Kichey et al., 2007) and is widely applied in Asian rice fields to achieve optimal rice production (Hameed et al., 2019). However, a major challenge is achieving a balance between increasing rice production and enhancing fertilizer use efficiency (Hameed et al., 2019). To address this challenge, numerous studies (e.g., Deng et al., 2014; Hameed et al., 2019; Hussain et al., 2022) have sought to identify optimal fertilization schedules that maximize N use efficiency in rice fields while minimizing negative impacts on soil and water resources and increasing rice yields. For example, in the upland rice areas of Thailand, Hussain et al. (2022) found that applying an optimal N fertilization rate of 90 kg N per hectare at the appropriate planting date (intermediate planting date: end of September or beginning of October) could increase rice yield and total rice N uptake by up to 105% and 159%, respectively, compared to conventional fertilization practices. This leads to better protection of the environment from nitrate contamination and increased economic benefits for farmers.

- Improved tillage practices

  Over the past few decades, intensive tillage practices such as deep plowing and lack of crop diversification in rice-growing areas of Asia have resulted in negative effects on soil properties, including reduced soil organic carbon (SOC) and decreased abundance of beneficial soil microorganisms (Hazra et al., 2014; Nandan et al., 2019). To mitigate these negative effects, conservation tillage practices such as zero-till, ridge-till, and mulch-till have been proposed to enhance soil properties (e.g., structure, bulk density, hydraulic conductivity, and aggregate stability), maintain sustainable rice production, and mitigate climate change (Nandan et al., 2019). For instance, in rice-based cropping systems of Patna, northeast India, Nandan et al. (2019) found that zero tillage practice combined with crop residue retention resulted in higher SOC, higher water stable macro-aggregates (thus increasing carbon sequestration capacity), and higher rice and wheat yields compared to conventional tillage without crop residue. However, it is important to note that the benefits of zero tillage practice are largely dependent on climate variables, particularly rainfall, and are more pronounced in drier areas (Chenu et al., 2019).

14.2.2 CROP ROTATION OPTIMIZATION

Compared to low-diversity crop production systems (e.g., only rice production), studies have shown that crop rotations with high crop diversity and biomass can significantly enhance crop yields across many regions of Asia (Ghosh et al., 2020; Han et al., 2020; He et al., 2021). This can be attributed to the fact that low-diversity crop production systems often lead to soil quality deterioration, unfavorable growth conditions for crops (e.g., increased pests and diseases), and higher production costs

(Garbelini et al., 2022). In contrast, crop rotations with high crop diversity can significantly enhance organic matter accumulation, promote nutrient cycling, increase water availability, improve soil properties, and protect soil from erosion and compaction issues (Garbelini et al., 2022). For example, in South Asia, Ghosh et al. (2020) found that incorporating grain legumes into lowland rice-wheat and upland maize-wheat rotations improved rice grain yield by up to 14%. In China, He et al. (2021) found that potato-rice rotation resulted in higher rice yields, better soil fertility, and higher economic profit compared to successive rice cropping. However, fertilizer application should be rationalized to maintain these benefits (He et al., 2021). Additionally, Han et al. (2020) showed that a double-cropping system of maize-rice rotation with incorporation of maize straw mulch and rice straw into the field increased annual yield by up to 15.2% compared to conventional cultivation without straw-return. In addition to its positive effects on agricultural production, crop rotation can also have a positive impact on agricultural GHG emissions (Hasukawa et al., 2021). For instance, in Japan, Hasukawa et al. (2021) showed that compared to continuous rice cultivation, the paddy-upland rotation system (i.e., a land-use system that rotates four crops: wheat-soybean-rice-rice over three years) reduced average agricultural GHG emissions (mainly CH_4 and N_2O) by about 5.6 Mg CO_2 eq. ha^{-1} y^{-1}, making this rotation system a powerful tool for mitigating climate change in Japan.

14.2.3 Enhancement of Crop Water Productivity

In addition to crop yield, CWP, defined as the ratio of crop yield to water consumed (Kijne et al., 2003), is also a fundamental indicator for evaluating the agricultural production level of a given cropland (Liu et al., 2021). In many regions of Asia, increasing CWP can be constrained by rainfall variability, water shortage, and unsuitable management of irrigation water (Letseku and Grové, 2022). Maximizing CWP while improving crop yields to relieve pressure on water resources and decrease conflicts in water use is currently a major challenge faced by the majority of Asian countries (Liu et al., 2021). There are many measures with great potential to raise CWP. Some of the major measures (Morita, 2021) are shown in Figure 14.3.

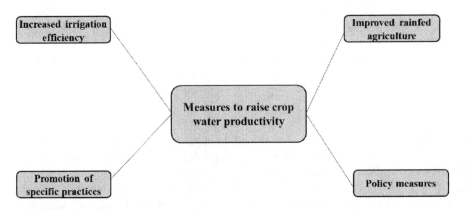

FIGURE 14.3   Measures to raise crop water productivity in Asia.

- Increased irrigation efficiency

  At a global level, it is estimated that only 50% of water (from rainfall or irrigation) is used by crops, meaning that 50% of the total available water is lost during storage, conveyance, and drainage after water application (Al-Faraj et al., 2016). In Asian agricultural croplands, water losses are much higher and can reach 60% in Central Asia due to poor irrigation networks and the practice of inefficient irrigation methods such as flood irrigation (Bekchanov et al., 2015; Sun et al., 2019). To avoid or at least reduce water losses in irrigated croplands, increasing irrigation efficiency is a promising measure (Sun et al., 2019). According to a report released by the International Bank in 2019, increasing irrigation efficiency (e.g., through irrigation modernization) across Central Asia has the potential to reduce water losses and improve crop yields by about 50% by 2050. In the same region, Sun et al. (2019) found that efficient irrigation methods (particularly sprinkle and drip) have the potential to significantly enhance irrigation efficiency and increase unit water benefit from 0.15 US\$m^{-3} to 0.24 US\$m^{-3}. Micro-irrigation methods could also be a promising option in South Asia (Mohan et al., 2022). However, there are currently some constraints that may limit the adoption of micro-irrigation technologies (Mohan et al., 2022). For instance, the small size of cultivated areas (0.5 to 3.0 ha, Thapa and Gaiha 2014), weak knowledge of farmers, and lack of financial and technical support (Chandio et al., 2020) could be major constraints (Mohan et al., 2022). Cooperative or collective farming could be an effective option to overcome these constraints (Mohan et al., 2022) as it can offer some benefits of large-scale farming such as decreasing the cost of cultivation and promoting the use of micro-irrigation technologies (Mohan et al., 2022).

- Promotion of specific practices

  Practices such as irrigation scheduling (i.e., regulating the amount and frequency of irrigation), regulated deficit irrigation, land leveling, water harvesting, and conservation tillage can significantly reduce water losses and increase crop yields (Shock et al., 2013; Aryal et al., 2020; Si et al., 2023). For example, Kang et al. (2002) found that spring wheat (*Triticum aestivum* L.) receiving reduced irrigation by 20% during early vegetative stages had grain yields equal to or even greater than wheat that was fully irrigated, leading to a significant decrease in the total irrigation water required over the full season length. This can be attributed to the fact that intensive irrigation may create permanent wetting conditions in the crop root zone, which can restrict oxygen supply to the roots, inhibit root growth, reduce water uptake by the roots, limit photosynthesis, and ultimately reduce crop yields (Liu et al., 2019; Si et al., 2023). Additionally, in Indian irrigated rice-wheat fields, laser land leveling has significantly decreased irrigation time and improved water-use efficiency (Jat et al., 2014; Aryal et al., 2015; Aryal et al., 2020). Specifically, it reduced irrigation time by up to 69 h ha^{-1} season^{-1} and saved irrigation water by up to 43% (Aryal et al., 2015, 2020; Jat et al., 2014).

- Improvement of rainfed agriculture

  Due to expected water shortages resulting from climate change and its associated negative effects on irrigation supply, farmers in rainfed crop production systems in Asia need to adopt more water-saving practices, improve
  the productivity of their rainfed lands, and make their agricultural activities
  more adaptive to climate change (Hayashi et al., 2018). For instance, in some
  states of India, farmers have adjusted their agricultural practices to rainfall variability by practicing conservation agriculture (Aryal et al., 2020).
  In Laos, Southeast Asia, compared to conventional agriculture, organic and
  clay-based soil amendments have increased CWP in rainfed production systems by up to 0.7 kg m^{-3}, making soil-based interventions promising options
  for enhancing agricultural productivity (Mekuria et al., 2015). Additionally,
  as runoff in Asian rainfed areas can contribute to up to about 50% of water
  losses (Anjum et al., 2010), practices adopted to decrease surface runoff
  such as effective rainwater harvesting, contour planting, land leveling, and
  forestation can significantly reduce soil and water losses and enhance moisture reserves in the soil (Baig et al., 2013). In addition to their key role in
  water conservation, these measures can also lead to crop yield enhancement.
  For example, Baig et al. (2013) reported that saving 1 mm of water in rainfed
  areas could increase wheat yield by an average of 10 kg ha^{-1}.
- Policy measures

  Policy measures such as consumption-based pricing of water and electricity
  and water/energy rationing are needed to raise awareness of water conservation, encourage farmers to waste less water and pollute less, invest more in
  water-saving practices, and encourage farmers to plant climate-resilient crops
  (Dhawan, 2017). These measures are also necessary to ensure water security
  and maintain food security in many Asian countries (Mu et al., 2019).

14.2.4 AGROFORESTRY

Over the past two decades, agroforestry, that is, the cultivation of agricultural crops
with trees and livestock on the same land, has gained significant attention in Asia
due to increased awareness of its benefits (Lin et al., 2021). It covers about 23%, 27%,
50%, and 77% of the total agricultural land in Northern and Central Asia, South
Asia, East Asia, and Southeast Asia, respectively (Lin et al., 2021). These high per-
centages can be attributed to the availability of many large lands suitable for agroforestry (Atreya et al., 2021). Advanced agroforestry policies and practices have been
established in some Asian countries (Lin et al., 2021), contributing to many benefits
such as increased carbon sequestration (thus increasing carbon stock in soil), soil
protection from extreme climate events, improved soil fertility, enhanced farmers'
income, and increased adaptation of crops to droughts, floods, and other natural dis-
turbances (Waldron et al., 2017; Aryal et al., 2020). For example, in degraded lands
in West Java, Indonesia, Siarudin et al. (2021) found that adopting a "mixed-tree
lots" agroforestry system (i.e., trees and natural undergrowth) resulted in a significant carbon stock estimated at 108.9 Mg ha^{-1}. The system also prevented soil erosion and helped restore degraded lands (Siarudin et al., 2021). In the same country,

Wardah et al. (2013) showed that simple agroforestry systems (combining individual trees and cash crops) in the Central Sulawesi region resulted in a carbon stock of up to 45 Mg C ha^{-1}, while complex agroforestry systems (combining multiple tree species, shrubs, bushes, and crops) had a carbon stock of up to 85 Mg C ha^{-1}.

In India, the planting of Eucalyptus and Populous trees in agricultural fields has been shown to double farmers' income due to benefits from fruits, timber, flowers, medicines, and other crops (Aryal et al., 2020). Similarly, in Indonesia, agroforestry may contribute up to five times more income (Duffy et al., 2021). In Nepal, people derived approximately 75% of their fuelwood from agroforestry systems (Khadka et al., 2021). Despite these promising results across some Asian countries, challenges such as lack of knowledge, institutional capacity, unclear policy support for agroforestry, limited extension services, and limited funding remain (Paudel et al., 2022). To enhance farmers' adoption of agroforestry, dynamic policies and measures backed by scientific research should be implemented. This includes promoting technologies to address the management complexity of agroforestry systems, providing institutional change, and generating funds for investment (Aryal et al., 2020; Atreya et al., 2021). Furthermore, more varied forms of agroforestry should be investigated throughout Asian countries with a focus on food security in relation to ongoing and future climate change (Duffy et al., 2021).

14.2.5   BIOFUEL PRODUCTION

As burning fossil fuels can emit a significant amount of CO_2 along with other GHGs (Saleem, 2022), renewable, sustainable, and environmentally friendly energy resources such as biofuels have gained increasing attention in Asia over the past few years as an efficient alternative to fossil fuels for decarbonizing the agriculture sector (Arai and Matsuda, 2018). For example, some biofuels such as bioethanol from sugarcane can reduce GHG emissions by up to 140% relative to gasoline (Arai and Matsuda, 2018). As crops with great potential for bioenergy production such as palm oil, maize, sugarcane, and coconut oil are widely cultivated in Asian cropping areas, biofuel production in Asia could be a promising option for mitigating climate change in the coming decades. However, it needs to be balanced with other social, ecological, and environmental concerns such as avoiding local biodiversity loss and providing new income for farmers (Acosta et al., 2016; Tudge et al., 2021). Achieving this balance could be a significant challenge.

14.2.6   SUSTAINABLE LIVESTOCK MANAGEMENT

Due to ongoing climate change, the majority of Asian countries face difficulties in sustaining livestock production (Nishanov, 2015). Over the coming decades, livestock production is likely to be adversely affected by climate change, particularly increasing temperatures (Rojas-Downing et al., 2017). Although livestock in Asia contributes significantly to GHG emissions, particularly CH_4 (up to 900 million tons CO_2 eq. in 2010; Otte et al., 2019), appropriate adaptation to climate change can offer opportunities for food security and a sustainable environment (Escarcha et al., 2018). Northern China, Mongolia, and South Korea have been identified as hotspots

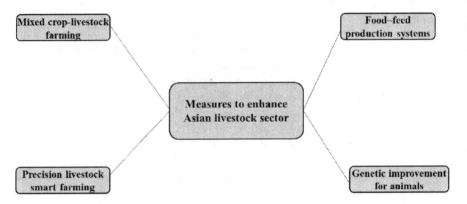

FIGURE 14.4 Measures to enhance Asian livestock sector.

for Asian and global livestock CH_4 emissions, indicating a need for increased mitigation efforts in these regions (Zhang et al., 2021). Researchers have identified several adaptation options that can increase the resilience of the Asian livestock sector to climate change while reducing GHG emissions and enhancing productivity (Escarcha et al., 2018). These options are presented in Figure 14.4.

*   Improved mixed crop-livestock farming:
    Mixed crop-livestock farming, which involves growing crops and raising livestock on the same farm, is an effective way to achieve food security without endangering the environment (Rojas-Downing et al., 2017). Mixed crop-livestock farming can contribute to efficient manure management by using livestock manure to fertilize crop farmlands and providing traction for farming (Danso-Abbeam et al., 2021). It can also improve food security, particularly in developing economies (Asante et al., 2018), decrease production costs (Asante et al., 2018), and reduce agricultural GHG emissions (Tang et al., 2019). For example, it has been estimated that mixed crop-livestock farming in China's Loess Plateau may reduce on-farm GHG emissions by up to 33%, making it a cost-effective option for mitigating climate change (Tang et al., 2019). To increase its resilience to climate change, mixed crop-livestock farming should be supported and improved through measures such as cultivating drought-tolerant crops, appropriate soil and nutrient management, and promoting agricultural technologies (Thornton and Herrero, 2014).
*   Precision livestock smart farming:
    Precision livestock smart farming involves the intensive use of modern agricultural technologies, such as sensors and computer algorithms, to gather accurate data about animal production on farms. This approach can improve animal health and welfare, optimize management practices, reduce on-farm labor and veterinary costs, enhance farm waste management, increase profits for farmers, improve environmental sustainability, and increase overall farm performance (Tzanidakis et al., 2023). However, there are currently

some constraints that may limit the adoption of precision livestock smart farming on a wider scale. These include the small size of farms in many Asian countries, lack of skills in using modern technologies, and limited knowledge diffusion (Guntoro et al., 2019; Skobelev et al., 2019). Addressing these constraints should be a priority.

- Food–feed production systems:
  Food-feed production systems integrate the leaves and by-products of on-farm food production into grass production for ruminant feeding (Halmemies-Beauchet-Filleau et al., 2018). In Asia, where agroforestry is relatively common (Lin et al., 2021), the leaves of fodder trees and shrubs, such as cassava, can provide a valuable protein supplement for ruminants. This can significantly enhance animal performance sustainably on small-holder farms (Halmemies-Beauchet-Filleau et al., 2018).

- Genetic improvement for animals:
  Genetic improvement for animals, along with a range of other measures and tools, can promote animal conservation, ensure efficient utilization of animal potential, and provide additional benefits for society. These benefits include healthier and more resilient livestock and reduced agricultural GHG emissions (Pineda et al., 2021; Wray-Cahen et al., 2022). However, further efforts are urgently needed to maximize the adaptability of this measure in some Asian countries, such as those in Southeast Asia, which are characterized by harsher climates and weaker adaptation capacity to climate change (Pineda et al., 2021).

14.3   CONCLUSIONS

Due to the high vulnerability of Asian agriculture to climate change, there is an urgent need to implement proper adaptation measures to sustain agricultural productivity, decrease vulnerability, and improve the resilience of agricultural systems to climate change in the coming decades. Although this chapter has identified many effective adaptation measures, such as better management of rice fields, crop rotation optimization, and agroforestry, there are still challenges and constraints in successfully integrating these measures within Asian agricultural systems. Successful implementation requires organizational and institutional support, particularly continuous funding from international institutions. Many ongoing adaptation measures in Asia are financed by international institutions such as the International Fund for Agricultural Development, the World Bank, and the International Financial Institution. However, changes in donor priorities can constrain their sustainability. These international institutions, together with Asian institutions, should ensure continuous funding for effective adaptation strategies that can help Asia reduce its agricultural GHG emissions and achieve the 1.5°C temperature drop required to fight climate change and associated threats to sustainable agriculture. Without sufficient funding, large-scale increases in agricultural production in Asia without damaging the environment may not be realistic.

15 Building a Sustainable Agroecological Environment

Insights into Agricultural Adaptation Options in the United States and Canada

15.1 INTRODUCTION

North America, particularly the United States and Canada, is one of the largest agricultural producers in the world (USDA-FAS, 2015; USDA-FAS, 2021). For instance, about a third of the global maize (*Zea mays* L.) production occurs in the United States, reflecting the country's crucial role in ensuring global food security as maize is a fundamental food crop for billions of people (USDA-FAS, 2015; Roesch-McNally et al., 2017). Additionally, about 20 million tons of Canadian wheat (*T. aestivum*) are exported yearly to many countries throughout the world, making a significant contribution to global food security (USDA-FAS, 2021). Despite these data showing the productivity of both countries, climate change over the next few decades is expected to have negative effects on agricultural production in both countries (Melillo et al., 2014; Roesch-McNally et al., 2017; Adekanmbi et al., 2023). Overall, climate change is projected to affect agricultural production through decreased yields of some crops and changes in production and quality of livestock (Hatfield et al., 2014; Roesch-McNally et al., 2017; Adekanmbi et al., 2023). For example, in response to climate change by 2050, maize yield in Texas, United States could reduce by up to 31% relative to the 1976–2005 baseline period (Kothari et al., 2022) while in Canada, climate change may lead to an 18% decline in potato yield in many areas by 2050 compared to the 1995–2014 baseline period (Adekanmbi et al., 2023).

To reduce potential effects and avoid risks related to current and projected changes in climate, farmers throughout both countries are currently practicing agronomic adaptation measures such as crop rotation, conservation agriculture, and integrated pest management (Gowda et al., 2018; Zhao et al., 2023). However, future effects of climate change on agricultural production in the United States and Canada will depend mainly on the rate and severity of change which are closely related to future greenhouse gas (GHG) emissions and the degree to which farmers can adapt (Gowda et al., 2018; Zhao et al., 2023). Therefore, future agricultural

DOI: 10.1201/9781003404194-18

activities in both countries require a transition from commercial agriculture to more sustainable agroecological farming that focuses on implementing environmentally friendly agricultural options that can simultaneously promote productivity and restrain further agricultural GHG emissions (i.e., sustainable agricultural options) (Roesch-McNally et al., 2017). A successful transition is essential for stabilizing agricultural production, ensuring global food security, and achieving the climate neutrality target by 2050 (Zhao et al., 2023). In this context, the main objective of this chapter is to analyze sustainable agricultural options one by one, especially their strengths, potential constraints that may hinder successful implementation, and recommendations for better agricultural performance under climate change by 2050.

15.2 AGRICULTURAL ADAPTATION OPTIONS IN THE UNITED STATES

Over the coming decades, agriculture in the United States is expected to face more extreme climate events, including droughts, increased growing season temperature (Wehner et al., 2017), and increased soil erosion due to intense rainfall events (Gowda et al., 2018), leading to decreased crop yields (Rosenzweig et al., 2014). As the United States is one of the top producers of food crops in the world, negative effects of climate change on agricultural production can damage both national and global food security (Ishtiaque, 2023). Therefore, it is essential to ensure that the United States adopts sustainable adaptation options that can simultaneously increase agricultural productivity and restrain further agricultural GHG emissions (Ishtiaque, 2023). This section summarizes those sustainable options one by one (Figure 15.1).

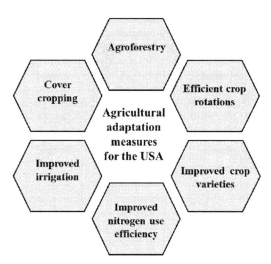

FIGURE 15.1   Climate change adaptation measures for Agricultural production in the United States.

15.2.1 Cover Cropping

Cover cropping, that is, the introduction of a cover crop between two main crops, is a promising and sustainable agronomic practice that can enhance both soil health and crop performance due to its high potential to decrease nitrate (NO_3^-) leaching from agricultural fields by scavenging residual nutrients, storing them in the soil, and making them available for future crops (Scavo et al., 2022; Gupta et al., 2023; Schnitkey et al., 2023). More precisely, this practice has a high potential to reduce the intensity of soil erosion, improve carbon sequestration, increase biodiversity, reduce nutrient leaching by up to 69% (Nouri et al., 2022), improve nutrient availability for crops, limit pests, regulate water infiltration, and improve the quality of drainage water (i.e., the introduction of a cover cropping between two main crops helps catch soil mineral nitrogen (N) before the period of drainage and consequently decreases NO_3^- leaching and NO_3^- concentration in the drainage water) (e.g., Blanco-Canqui et al. 2015; Justes, 2017; Adetunji et al. 2020) (Table 15.1). Furthermore, cover cropping is one of the best management practices (BMPs) for decreasing NO_3^- leaching and enhancing soil resilience in croplands facing growing concerns of increasing groundwater NO_3^- concentration in their wells (Iqbal et al., 2021). For instance, in the Waverly Wellhead Protection Area, Nebraska, Midwestern USA, Iqbal et al. (2021) showed that growing a cereal rye cover crop during the fallow winter period can reduce soil NO_3^- concentration by up to 20.4 Kg-N ha^{-1} relative to soils without cover crops. However, Gupta et al. (2023) showed that cover cropping should be supported by other robust and cost-effective farm management practices (e.g., proper fertilizer management) to aid cover cropping benefits in preventing nutrient loss from agricultural fields. Moreover, overall, the cover cropping practice can greatly reduce GHG emissions by up to 150 g CO_2 eq. m^{-2} y^{-1} (Kaye and Quemada, 2017). In Pennsylvania, United States, cover cropping practice has reduced GHG emissions by up to 46 g CO_2 eq. m^{-2} y^{-1} (Kaye and Quemada, 2017). On the other hand, by applying the optimal fertilizer amount to the main crop and trapping post-harvest nutrients, cover cropping can also significantly improve crop yields (Bourgeois et al., 2022; Van Eerd et al., 2023). For instance, Van Eerd et al. (2023) reported that maize yield increases with leguminous cover crops ranged between 7 and 33%, with cover crop mixtures ranging from 22% to 30%.

TABLE 15.1
Advantages of Using Cover Crops

Advantages	Reference
Improve soil quality	Haruna et al. (2020)
Increase soil organic matter	Koudahe et al. (2022)
Improve soil microbial diversity	Vukicevich et al. (2016)
Reduce soil and water losses	Cerdà et al. (2022)
Enhance plant nutritional status	Scavo et al. (2022)
Decrease nitrate leaching to groundwater	Nouri et al. (2022)
Improve ecosystem multifunctionality	Restovich et al. (2022)

15.2.2 Improved Crop Varieties

Crop improvement through modern breeding techniques (e.g., modifying the expression of genes involved in climate stress responses) is an excellent way to both enhance crop yields and reduce agricultural GHG emissions (Razzaq et al., 2021). As an alternative to inorganic fertilizers that can lead to air and water (groundwater) pollution, the development of improved crop varieties with reduced reliance on N could also significantly decrease GHG emissions, particularly nitrous oxide (N_2O) emissions, reducing air and water (groundwater) pollution (Khalil and Osborne, 2022). Different improved crop varieties are currently being cultivated by farmers across the United States, helping them produce more food despite climate change (Cooper and Messina, 2023). While many challenges persist, successful outcomes from a range of improved crop varieties such as maize (e.g., Gaffney et al., 2015) and wheat (Guarin et al., 2022) have been recorded in various environments across the United States (e.g., water-limited environment, saline environment) (Cooper and Messina, 2023). These positive outcomes could be explained by the fact that improved crop varieties have great potential to limit water loss, enhance water uptake, enhance osmotic adjustment, and tolerate new abiotic and biotic pressures (Varshney et al., 2021). Moreover, studies (e.g., Varshney et al., 2021; Diepenbrock et al., 2022) have estimated that improved crop varieties may hold the key to addressing the climate change challenge in the agriculture sector in the United States and providing resilient agricultural production systems for the coming decades, particularly by 2050. Therefore, in the coming years, there is a great scope to maximize yields and reduce GHG emissions in different environments across the United States by enhancing crop breeding techniques to improve upon yield-advancing physiological traits regardless of climate change scenarios and deploying appropriate crop varieties along with other fundamental sustainable adaptation options associated with climate change resistance (Guarin et al., 2022).

15.2.3 Improved Irrigation

Due to the potential increase in moisture deficit as a direct result of rising temperatures and severe drought events, it is expected that crops, especially in Arizona, California, Montana, Nevada, and Idaho, will require more water in the coming decades to reach yield targets. For instance, compared to 2014, maize in the Midwest is expected to need about 35% more water inputs to reach yield targets by 2050 (Lobell et al., 2014; Ort and Long, 2014; Basso et al., 2021). Furthermore, climate change in the United States in the coming decades will lead to an increase in the proportion of agricultural land under irrigation practice. McDonald and Girvetz (2013) expected that relative to the baseline 1985–2005 period, the total area irrigated will increase by 4.5 to 21.9 million ha by 2030, depending on GHG emissions scenarios. To cope with these expectations, there is no option other than improving irrigation efficiency on-farms via BMPs (McDonald and Girvetz, 2013). Otherwise, in the absence of gains in irrigation water-use efficiency, total water withdrawals for irrigation may increase substantially, leading to water insecurity by 2050 (McDonald and Girvetz, 2013). Enhancing irrigation efficiency is considered a promising response to climate change, especially in arid areas (Frisvold et al., 2018). It can allow farmers to reduce irrigation water amount and increase crop yields, thus decreasing costs of irrigation and increasing profits (Frisvold et al., 2018). In Louisiana, Arkansas, Arizona, Wyoming, and Colorado states where about 90% of croplands are irrigated (Gua Mpanga and Idowu, 2021), moving away from surface irrigation has great potential to generate substantial irrigation water-use efficiency gains, offsetting demand for an increase in irrigation rate in these states (McDonald and Girvetz, 2013). Also, Malek et al. (2018) indicated that investing in a more efficient irrigation system can improve the agricultural economy of the Yakima River Basin in Washington by up to 25%. Furthermore, Genius et al. (2014) showed that developing intensive extension and educational programs on irrigation could provide producers with adequate information to transition from unsustainable traditional irrigation practices to smart irrigation technology that may lead to water saving in agriculture (Gua Mpanga and Idowu, 2021).

Moreover, Gua Mpanga and Idowu (2021) revealed that improved irrigation technologies and the use of best management decisions (e.g., installation of soil moisture sensors, use of drip irrigation systems instead of surface irrigation, use of plant sensors) in Arizona state over 2007–2017 have resulted in a reduction in the quantity of irrigation water used by about 5%. Considering that currently only about 20% of farms in Arizona gather information from more advanced technologies when making irrigation decisions, further reduction could be gained over the coming years by promoting advanced technologies among farmers (Frisvold et al., 2018). Encouraging farmers to use local climate and soil information is helpful for meeting crop needs without water loss and irrigating more precisely (Evett et al., 2020). On the other hand, in the US Great Plains, one of the world's largest food production areas, automated deficit irrigation treatments in the range of 50% to 80% of full irrigation have shown crop yields close to or even exceeding those of full irrigation, thus improving on-farm irrigation efficiency and producing greater crop water productivity (Evett et al., 2020). Remote sensing technologies in the United States also provide a great opportunity to quantify crop water use (Evett et al., 2020). A system developed by the Agricultural Research Service of the United States Department of Agriculture (USDA-ARS) and the National Aeronautics and Space Administration (NASA) can provide daily evapotranspiration with resolutions as fine as 30 m, making it a current and future powerful tool to estimate water use efficiency over a large area (Evett et al., 2020). Finally, it is worth mentioning that improved irrigation together with improved crop varieties have great potential to make irrigated agriculture in the United States more economically sustainable with decreasing water supply (Evett et al., 2020).

15.2.4 AGROFORESTRY

Agroforestry, that is, the integration of trees, shrubs, and livestock into croplands, is a promising land-use option with great potential to address global climate change as it highly promotes the carbon sequestration process (i.e., capturing and storing carbon in soils and tree biomass), with the possibility of increasing crop yields (Jose and Bardhan, 2012; Kim et al., 2016; Castle et al., 2022). Furthermore, it offers many other environmental, economic, and social benefits such as providing sustainable production, improving food security (via higher crop yields), enhancing water quality, and improving biodiversity (Udawatta and Jose, 2012; Wilson and Lovell, 2016; Castle et al., 2022). Moreover, it offers a great opportunity to advance the United Nations 2030 Sustainable Development Goals (SDGs) (Waldron et al., 2017). Relative to conventional agriculture, agroforestry systems (e.g., silvopasture, alley cropping, windbreaks, riparian buffers, and forest farming) can decrease surface runoff, soil erosion, organic carbon, and related nutrient losses by about 58%, 65%, 9%, and 50%, respectively, while lowering herbicide, pesticide, and other pollutant losses by about 49% (Chenyang et al., 2021). In addition to that, agroforestry usually leads to less fertilizer application as livestock manure may be used on-farm (Chenyang et al., 2021). Given these diverse potential benefits, agroforestry across the United States has gained some attention from farmers over the past years (Castle et al., 2022; Smith et al., 2022). Due to high resource availability (extensive forest land) and diverse ecology, agroforestry practice has shown good success in some regions across the United States during the past years (Baker and Saha, 2018). For instance, in 2012 incorporating trees on 10% of all US agricultural land resulted in good carbon sequestration estimated at 530 MMT CO_2 per year (Chenyang et al., 2021). An earlier study investigated soil quality and growth of southern pine trees in southern USA and found that silvopasture system (from existing woodlands by removing non-pine vegetation, thinning pine trees, and planting suitable cool-and warm-season forages) is showing a better environment for faster growth of southern pine trees versus woodlands. Despite these promising findings, constraints and issues still remain (Chizmar et al., 2022). There is an urgent need to expand agroforestry across the United States as in 2017 this land-use option was adopted on only 1.5% of total agricultural land with adopters concentrated in the mid-Atlantic and Pacific Northwest regions (Chenyang et al., 2021). Also integrating agroforestry into croplands requires considerable planning and technical expertise to ensure that system components complement one another and do not compete (Smith et al., 2022). Furthermore, further research education and outreach are required to support informed and appropriate adoption of agroforestry in the United States (Smith et al., 2022). Overall, even though the United States has one of the major agroforestry policies in the world the USDA Agroforestry Strategic Framework Plan (FY 2011–2016) (USDA, 2011) robust agricultural policy reform must be rooted in a clear understanding of the constraints farmers face in adopting and sustaining agroforestry practices across various farms of the United States (Chenyang et al., 2021).

15.2.5   Improved Nitrogen Use Efficiency (NUE)

Agriculture in the United States cycles large amounts of N to produce food, fuel, and fiber (Robertson et al., 2013). This usually results in huge amounts of reactive N in agricultural systems which can lead to environmental pollution, e.g., groundwater pollution by N (Ward et al., 2023). As fertilizer N use in the country is expected to double from the average 2010–2016 level by 2030 (Robertson et al., 2013), sustainable N management practices are needed to reduce N losses in agricultural systems and their associated effects on the environment (Cormier et al., 2016). There are many ways

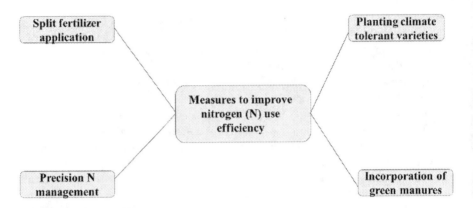

FIGURE 15.2   Measures to improve nitrogen (N) use efficiency.

(Figure 15.2) to improve N use efficiency (NUE) (Ward et al., 2023). Split fertilizer application (i.e., applying fertilizer in two or more applications), precision N management (e.g., via use of machine learning), and incorporation of green manures into soil have emerged during the past years as good ways to synchronize N supply with crop demand (Carter et al. 2014; Finney et al. 2015; Rees et al., 2020; Ward et al., 2023). For instance, in Verona, Wisconsin, United States, Zhou and Soldat (2022) revealed the good ability of a machine learning-based turf growth model using the random forest algorithm (ML-RF model) to enhance creeping bentgrass (*Agrostis stolonifera*) putting green management by estimating short-term clipping yield. Also, contrary to traditional uniform N fertilizer application that neglects within-field spatial variability, variable rate technology nitrogen (VRT-N) applies fertilizers at different rates to meet site-specific demand and can reduce the risks of over- and under-fertilizing US maize crops (Jin et al., 2019). Despite the availability of effective ways and tools that can help improve NUE within US agricultural systems, it is worth mentioning that further research on agricultural N cycle responses to changing climate and development of quantitative models capable of predicting N-climate interactions with confidence across a wide variety of crop-soil-climate combinations are needed to ensure sustainable N management under climate change (Robertson et al., 2013). Moreover, transitioning from chemical fertilizers to organic agriculture could be helpful for getting healthy soils with active biological activity and good physical and chemical soil properties (Rodale Institute, 2012; Altieri et al., 2015). For instance, Altieri et al. (2015) reported that transitioning from chemical to organic maize agriculture in the United States since 1981 has resulted in an increase in maize yields by about 31%, especially during years of droughts. These drought yields are notable when compared to genetically engineered "drought-tolerant" varieties which saw increases of only 6.7 to 13.3% over conventional (non-drought resistant) varieties (Altieri et al., 2015).

15.2.6   Efficient Crop Rotations

Compared to monocropping, efficient crop rotation (i.e., the practice of growing different crops in succession on the same land) can show better crop yields and quality, provide higher revenue for farmers, and lower production risks (e.g., disease risks) (Khakbazan et al., 2022). Crop rotation in the United States is practiced over the majority of croplands (Wallander, 2020). According to the USDA's Agricultural Resource Management Survey, only 18% of cropland planted to maize in 2010 was in maize continuously over the previous three years (Wallander, 2020). For wheat, the numbers are similar; only 14% of spring wheat acreage in 2009 was in wheat continuously over the previous three years (Wallander, 2020). For soybeans, continuous cropping is less common (Wallander, 2020). Rotating cash crops such as maize, wheat, and soybeans with cover crops including legumes, grasses, and forbs has resulted in several benefits for both crops (improvement of yields) and soils (enhancement of soil properties), leading to notable economic profitability (Cai et al., 2019; Bowles et al., 2020). For instance, Bowles et al. (2020) investigated crop rotation diversification effect on agricultural outputs over all the United States and Canada and showed that more diverse rotations have improved maize yields over time and across all growing conditions by about 28%. They also showed that even under unfavorable conditions (e.g., in drought years), more diverse rotations can reduce maize yield losses by 14% to 89.9% relative to monocropping. This positive effect of diversified rotations on maize yield under drought and other types of stressful growing conditions could be attributed to enhanced soil properties such as increases in soil water capture and storage and abundance of beneficial soil microbes (Bowles et al., 2020). In alignment with these findings, in Canada Gaudin et al. (2015) also found 7% higher maize yields during hot and dry years in a diversified five-crop rotation than in a maize-soybean rotation.

15.3 AGRICULTURAL ADAPTATION OPTIONS IN CANADA

Over the coming decades, it is expected that agriculture in some regions across Canada will face more extreme climate events, especially droughts, which may lead to decreased yields of some crops (Leng and Hall, 2019). However, in other regions, climate change may lead to some beneficial changes (e.g., higher level of pulse production, increase in the area under cultivation and its productivity), but if the rates of these changes are faster than farmers have faced, they may result in more difficulties for adaptation (Kulshreshtha, 2019). For instance, cultivation of new crops may become economically feasible (Kulshreshtha, 2019). Therefore, it is essential to ensure that farmers in Canada are aware of these effects of climate change and are adopting sustainable adaptation options that can simultaneously increase agricultural productivity and restrain further agricultural GHG emissions (Smit and Skinner, 2002; Kulshreshtha, 2019). This section summarizes those sustainable options one by one (Figure 15.3).

15.3.1 AGROFORESTRY

Over the past two decades, agroforestry has gained attention from farmers due to provincial programs, support from government and nongovernment organizations, and increasing demand for sustainable agricultural practices (Isaac et al., 2018; Ma et al., 2022). This has led to the implementation of various agroforestry systems such as

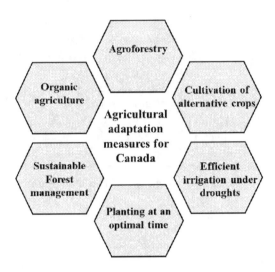

FIGURE 15.3 Climate change adaptation measures for Agricultural production in Canada.

integrated riparian systems, intercropping, windbreaks, forest farming, and silvopas-
toral systems (Thevathasan et al., 2012; Ma et al., 2022). These systems have resulted
in numerous agronomic, environmental, economic, and social benefits including
enhanced agricultural productivity, biodiversity conservation, income diversification, and carbon sequestration (Thevathasan et al., 2012; An et al., 2022; Ma et al.,
2022). For instance, in the Prairie, Shahariar et al. (2021) found that riparian buffer
systems have resulted in higher abundance of soil organic carbon (SOC) in subsoil
with good potential for C sequestration relative to annual crop and pasture. Also, in
southern Ontario, Ofosu et al. (2021) investigated SOC sequestration with different riparian systems and revealed that tree buffers have shown better potential for
SOC sequestration than grass buffers. Moreover, in central Alberta, Ma et al. (2022)
assessed the impact of three different agroforestry systems (hedgerow, shelterbelt,
and silvopasture) and their component land-use types (treed areas and adjacent
cropland/grassland) on total C stock in soil. The results revealed that the treed area
within agroforestry systems stored about 178.5 t ha^{-1} more C compared to adjacent
cropland or grassland. Despite these promising outcomes of agroforestry across
Canada, it is worth mentioning that some challenges and constraints still exist and
should be addressed to ensure high performance of agroforestry systems (Thevathasan
et al., 2012; Ma et al., 2022). For instance, some regions in the country (e.g., the
Atlantic Region) lack technical support and expertise to advise landowners about the
concepts and principles of agroforestry and how to incorporate these into their croplands (Thevathasan et al., 2012). Therefore, training and workshops by agroforestry
experts could be helpful to overcome this lack (Laforge et al., 2021). Also, quantitative information on C stocks from particular agroforestry systems is still lacking (Ma
et al., 2022). Thus, further research on this topic is highly needed. Moreover, there
is a need to maintain current agroforestry systems across Canada as their C stocks
increase over time (Ma et al., 2022).

15.3.2 ALTERNATIVE CROPS

One of the major effects of climate change on the agriculture sector is that crops which were once the most suitable in a specific region may no longer be the suitable crop for that region (Mu et al., 2023). This critical change in crop suitability is mainly attributed to changes in climatic needs of the crop, length of growing season, resilience to climate extremes, and vulnerability to pests and diseases with expanding ranges (Mu et al., 2023). For example, because of warmer and longer growing seasons, soybean cultivation may shift northward into Saskatchewan and other parts of the Prairie region (Kulshreshtha, 2011; Campbell et al., 2014) and climate may become more suitable for maize cultivation on the prairies (Campbell et al., 2014). Also, sorghum could be well suited to new climate as it can develop an extensive root system early and has great resistance to water deficit conditions (Almaraz et al., 2009; Campbell et al., 2014). Furthermore, an expected increase in the number of degree days in the valley areas surrounding Prince George by the 2050s would be an optimal climatic condition for cultivating canola and other crops that previously could not have been cultivated in this area (Picketts et al., 2009; Campbell et al., 2014). Canadian farmers can adapt to this critical change by cultivating alternative crops that are more resilient to variations in temperature and precipitation (Mu et al., 2023). For instance, in the Cariboo region of British Columbia, most crops are cultivated to feed animals. So as long as the new crop is palatable to the animals and shows similar nutritional value there would be an interest in trying new varieties (Mu et al., 2023).

15.3.3 EFFICIENT IRRIGATION UNDER DROUGHTS

Drought events can highly increase evapotranspiration and rapidly deplete soil water, inducing water stress and limiting yield potential, as is the case with spring wheat in Saskatchewan, Canada (He et al., 2012; Wu et al., 2023). Therefore, for regions that should expect more droughts during the next years (e.g., the Prairies), efficient irrigation together with reducing soil evaporation could be one of the best adaptive options under drought conditions (Sauchyn and Kulshreshtha, 2010; Kulshreshtha, 2019). They have great potential to avoid water stress in soils, stabilize rainfed and irrigated crop yields over drought events, increase agricultural production, and support mixed farms in maintaining livestock activities via forage production (Brown, 2017; Kulshreshtha, 2019). For instance, in Saskatchewan, Canada, Wu et al. (2023) found that reducing soil evaporation by 50% brings synergetic benefits to the overall water-energy-food nexus by enhancing rainfed crop yield and water productivity, decreasing water use and energy consumption for irrigation, and improving hydropower production.

15.3.4 SUSTAINABLE FOREST MANAGEMENT

According to the Long-Term Low Emissions and Development Strategy (LT-LEDS), Canada has committed to decreasing its GHG emissions by 80% by 2050 relative to 2005 emissions (Environment and Climate Change Canada, 2016; Zerriffi et al., 2023). Due to their high potential to remove carbon from the atmosphere and store it, Canadian forests can take part in making this 2050 goal possible (Zerriffi et al., 2023). However, it is worth mentioning that climate change in Canada may pose a number of challenges to sustainable forest management, particularly challenges in how to reduce risks and increase opportunities that climate change poses for Canada's forests (Edwards and Hirsch, 2012). Effective management of Canadian forests could be ensured through a range of measures such as wildfire prevention, creation of a mix of tree species types and ages, rehabilitation of degraded forests, promotion of sustainable management practices on private woodlots together with other proactive options for avoiding potential irreversible impacts such as species extinctions (Lemieux et al., 2010; Lemieux et al., 2011). Finally, it is worth indicating that even though Canada has some of the most stringent forest management legal and policy frameworks (Gilani and Innes, 2020), the implementation of successful adaptation measures for Canadian forests always needs the participation of the widest possible range of stakeholders throughout the country such as scientists, policymakers, corporate leaders, resource managers, and forest practitioners (Laatsch and Ma, 2015; Ameztegui et al., 2018).

15.3.5 OPTIMAL CROP PLANTING DATE

Crop planting date is a fundamental factor affecting the growth and productivity of crops (Dong et al., 2019). Planting crops at an optimal time has good potential to avoid the negative effects of unsuitable climate conditions on crop growth, leading to crop yield enhancement (May et al., 2004; Dong et al., 2019). For instance, in southern Ontario, Page et al. (2021) showed that the first two weeks of September is the optimal seeding period for winter canola as this period is favorable for reducing winterkill to about 20%, maximizing yield potential, and ensuring optimal oil quality.

15.3.6 ORGANIC AGRICULTURE

Over the past three decades, organic farming practices such as crop rotations, intercropping, cover crops, use of organic fertilizers, composting, biological pest control, and minimum tillage together with other factors (e.g., extension of croplands) have significantly improved soil fertility, increased C sink, reduced soil degradation (e.g., erosion), enhanced soil biodiversity, reduced nutrient losses, and created more stable soil conditions for crop growth (Fan et al., 2019; Lychuk et al., 2019). Total C input to soils from crops, manure, and biosolids in Canada increased by 81% between 1971 and 2015, which shifted Canadian agricultural lands from a CO_2 source before 1990 (-1.1 Tg C yr^{-1}) to a small sink during 1990–2005 (4.6 Tg C yr^{-1}), and a larger sink thereafter (10.6 Tg C yr^{-1}) (Fan et al., 2019). Many Canadian farmers have currently taken big steps toward organic farming practices on their farms because it is economically and financially beneficial to do so (Laforge et al., 2021). However further encouragement and support of farmers are always needed.

15.4 CONCLUSIONS

Over the coming decades, it is expected that climate change may affect the productivity of the top agricultural producing countries in the world, which means that sustainable adaptation measures are urgently needed. Due to their important contribution to global food production, a successful transition from conventional agriculture to sustainable agriculture practices in the United States and Canada over the coming decades is fundamental to mitigate and adapt to climate change and meet growing food demands. This chapter reviewed these sustainable agriculture practices and found that the capacity of the agriculture sector in both countries to adapt to climate change is currently being strengthened by many sustainable measures such as cover cropping, agroforestry, and organic agriculture. However, these suitable climate change adaptations need to be more rigorously assessed to understand their wider strengths and limitations. Moreover, since climate change may create some opportunities for some regions across Canada, updating farmers about these opportunities is essential to maximize crop yields and minimize time, costs, and labor.

General Conclusions

This book presents five essential conclusions. First, even under optimistic scenarios, the global average temperature is projected to increase by approximately 1.5 to 2°C by 2050. This will result in heavy precipitation events becoming 1.7 times more frequent and 14% more intense. Global sea levels will also continue to rise by an additional 15 to 43 mm by 2050, depending on greenhouse gas emissions scenarios. This could increase the frequency and severity of coastal flooding events and associated fatalities in several regions worldwide, particularly in coastal areas of South Asia, North America, and East Sub-Saharan Africa. Droughts and heatwaves are also projected to affect over 75% of the world's population by 2050. Collectively, these changes pose significant challenges for achieving sufficient global agricultural productivity and food security by 2050. Second, achieving global climate neutrality and sufficient global agricultural productivity targets by 2050 is still behind schedule. Countries, particularly top emitters such as China, the United States, and the European countries, should take immediate action to reduce emissions and boost their environmental and agricultural measures and policies. If actions are not taken quickly to curb greenhouse gas emissions, agricultural productivity could fall dramatically by 2050, leading to catastrophic food insecurity in many countries, especially developing countries. This may result in millions of deaths due to hunger. Third, climate change in 2050 is likely to have more severe effects on agriculture in developing countries than in industrial countries. Severe crop yield and livestock losses are expected to occur in Africa and Asia. Policymakers should provide financial and technical support to these regions as soon as possible before they are adversely and irreversibly affected. Fourth, climate change by 2050 may create opportunities for some countries (e.g., Canada, Northern Europe). Updating farmers about these opportunities is essential to maximize crop yields and minimize time, costs, and labor. Finally, many sustainable agriculture measures such as cover cropping, cultivation of improved crop varieties, agroforestry, improved nitrogen use efficiency, efficient crop rotations, promotion of new crops, efficient irrigation under droughts, cultivation under optimal conditions, organic agriculture, and sustainable forest management could help mitigate the negative effects of climate change on agriculture by 2050. However, to maximize the benefits of these measures, close collaboration between farmers, practitioners, policymakers, and researchers is highly needed.

General Recommendations (50 Essential Recommendations)

1. One of the primary challenges facing humanity in the coming years is the production of sufficient food for a growing population amid a changing climate and environmental degradation. A focus on sustainable agricultural practices, often referred to as an *"Agricultural Revolution"* may provide a solution to this challenge.
2. Limiting global warming to 1.5°C by 2050 is crucial to mitigate the most severe impacts of climate change. This goal cannot be achieved without reducing greenhouse gas emissions from agriculture and implementing sustainable agricultural practices.
3. As major emitters, China, the United States, and the European Union must significantly reduce their greenhouse gas emissions within the next 27 years. Failure to do so would make it impossible to achieve global food security and climate neutrality by 2050.
4. The primary challenge in the coming years is to balance maximizing farm productivity with preserving environmental benefits. The international community must successfully address this challenge to achieve climate neutrality and zero hunger by 2050.
5. Climate change may create opportunities for crop cultivation in new agricultural regions in some countries, such as Canada and Finland. Farmers and decision-makers should be aware of these opportunities to maximize agricultural production.
6. The agriculture sector requires innovative options, measures, and approaches to adapt to the various issues and constraints faced by farmers, landowners, and their communities.
7. A transition from commercial to sustainable agroecological farming is necessary in each country. This involves implementing environmentally friendly agricultural practices that promote productivity while reducing greenhouse gas emissions.
8. Complex policy mixes are essential to avoid unexpected scenarios regarding global agricultural productivity under climate change by 2050.
9. Scientific efforts must focus on adapting crops to not only withstand extreme climate events but also to be productive under a wider range of climate conditions.
10. The development of water-saving agriculture is crucial for conserving water resources under climate change and for improving crop production to meet current and future demand by 2050.
11. If agriculture is practiced using scientific methods and improvement measures are adopted, sustainable crop and animal production can be achieved by 2050.

12. There is significant potential for increasing agricultural productivity world-wide. However, overcoming barriers and constraints is necessary to realize this potential.

13. Water pricing for the agriculture sector should be regularly reviewed and revised in light of expected climate change over the coming decades to promote water conservation.

14. Agricultural technologies that control greenhouse gas emissions, facilitate agricultural activities, and increase crop and animal production should be widely adopted and implemented.

15. Sustainable financing and subsidies for Africa are essential to ensure its contribution to global food security and climate neutrality by 2050.

16. Institutional, financial, and technical support is urgently needed for both Africa and Asia to increase the resilience of their agriculture to climate change over the coming decades.

17. Public-private partnerships can capitalize on the opportunities presented by agricultural adaptation measures to achieve national and international targets for climate change and agricultural production by 2050.

18. Replanting forests, restoring damaged agricultural lands, and diversifying crops have great potential for adapting to changing climates.

19. The development of innovative solutions to mitigate the negative effects of climate change on agriculture is urgently required.

20. Integrating agricultural adaptation measures into each country's development planning is necessary to achieve national and international targets for climate change and agricultural production by 2050.

21. Institutional change and flexibility are key challenges for enabling agricultural adaptation, as adaptive management approaches must be regularly reviewed to meet ongoing and future climate change.

22. There is an urgent need to reduce greenhouse gas emissions from Asian rice fields through proper measures such as transitioning to new rice varieties, optimizing fertilization, implementing water-saving practices, and improving tillage practices.

23. Research on the effects of climate change and adaptation in agricultural systems must move beyond regional contexts and consider key vulnerability priorities.

24. A thorough understanding of the specific properties of each agricultural area is necessary for the effective implementation of adaptation measures.

25. Future food security by 2050 will depend on the ability of staple crops to tolerate new abiotic and biotic pressures. Continuous development of improved crop varieties is therefore essential.

26. Addressing the ongoing and future effects of climate change on agriculture requires a sophisticated mix of policies and adaptation options rather than single-oriented solutions.

27. Agricultural adaptation options should be flexible and easily understood by farmers to maximize their positive outcomes.

28. In arid and semi-arid agricultural areas, irrigation will continue to be economically important for the coming decades. Sustaining irrigation is crucial for maintaining productivity in these areas.

29. Promoting local crop varieties that are resistant to climate change is highly recommended.
30. Agroforestry has a fundamental role in mitigating climate change. Further research is needed to determine the regional potential for developing agroforestry systems and to assess their potential for reducing greenhouse gas emissions at regional and national scales.
31. Sustainable nitrogen management practices in agricultural systems are urgently needed to reduce nitrogen losses and their associated environmental impacts.
32. Research has shown that agroforestry systems have great potential for increasing carbon storage in above- and below-ground biomass as well as in soil. Investing in agroforestry systems can therefore be a powerful option for meeting greenhouse gas reduction goals and mitigating climate change.
33. Farm practices should be regularly modified to adapt to changing climatic and non-climatic conditions, including climate variability and extremes as well as political, economic, and social factors. Decision makers should implement adaptation policies that take these factors into account.
34. Maintaining traditional crop varieties such as maize and rice and ensuring access to seeds is essential for adaptation by poor farmers in Asian and African countries.
35. The optimal adaptation option for maximizing crop yield and minimizing greenhouse gas emissions must be determined based on local conditions.
36. Without adequate financial and technical support for Africa and Asia, achieving global food security and climate neutrality by 2050 would be impossible.
37. Adapting the farming calendar to ongoing climate change is essential to avoid the negative effects of unsuitable climate conditions on crop growth.
38. Planting heat-resistant crops is highly recommended to avoid crop yield losses under extreme climate events such as heatwaves and droughts, thereby preventing income losses.
39. Investing in green infrastructure is highly recommended to improve agricultural production, reduce greenhouse gas emissions, control microclimate, enhance biodiversity, and improve human health.
40. Renewable energy sources such as wind, solar, and biomass should be intensively used on croplands in the coming years instead of unsustainable fuels to reduce agricultural greenhouse gas emissions. These renewable energy sources can provide farmers with additional jobs and income.
41. Soil conservation measures such as crop rotations, intercropping, cover crops, use of organic fertilizers, and minimum tillage are crucial for improving soil fertility, preventing soil degradation, enhancing soil biodiversity, reducing nutrient losses, and creating stable soil conditions for crop growth.
42. Organic agriculture should be promoted as a proactive option with high potential for producing food while establishing an ecological balance to prevent soil fertility or pest issues.
43. Close collaboration between policymakers, researchers, farmers, and practitioners is essential to maximize the benefits of any agricultural adaptation measure.

44. Soil carbon sequestration should be promoted as an effective solution for mitigating climate change over the next few decades.
45. Farmers should be regularly updated on new agricultural adaptation measures that can help them enhance farm productivity while reducing greenhouse gas emissions.
46. Further research is needed on the benefits of each adaptation measure to avoid unexpected negative effects.
47. The interaction of climate and crop growing conditions with proposed adaptation measures should be further studied through long-term experiments and modeling.
48. In addition to climate and soil conditions, prediction of crop yield under climate change must integrate aspects of crop development, physiology, biochemistry, and genetics.
49. Further work is needed to reduce uncertainty associated with modeling crop yields under future climate scenarios, particularly under extreme climate events.
50. As food security is threatened by land degradation processes such as soil erosion, salinity, and compaction, rehabilitation of degraded lands worldwide is urgently needed.

References

AAFC [Agriculture and Agri-Food Canada] 2011. An overview of the Canadian Agriculture and Agri-Food system 2011; Agriculture and Agri-Food Canada, publication 11279E.

Aalto, J, P Pirinen, PE Kauppi et al. 2022. High-resolution analysis of observed thermal growing season variability over northern Europe. Clim. Dyn., 58, 1477–1493. https://doi.org/10.1007/s00382-021-05970-y

Abara, M, K Komariah, S Budiastuti 2020. Drought frequency, severity, and duration monitoring based on climate change in Southern and Southeastern Ethiopia. IOP Conf. Ser. Earth Environ. Sci., 477, 012011. https://doi.org/10.1088/1755-1315/477/1/012011

Abate, TM, TM Mekie, AB Dessie 2022. Analysis of speed of improved maize (BH-540) variety adoption among smallholder farmers in Northwestern Ethiopia: Count outcome model. Heliyon, 8(10), e10916. https://doi.org/10.1016/j.heliyon.2022.e10916

Abbas, G, S Ahmad, A Ahmad et al. 2017. Quantification the impacts of climate change and crop management on phenology of maize-based cropping system in Punjab, Pakistan. Agric. For. Meteorol., 247, 42–55.

Abbass, K, MZ Qasim, H Song et al. 2022. A review of the global climate change impacts, adaptation, and sustainable mitigation measures. Environ. Sci. Pollut. Res., 29, 42539–42559.

Abdallah, L, T El-Shennawy 2013. Reducing carbon dioxide emissions from electricity sector using smart electric grid applications. J. Eng., 845051, https://doi.org/10.1155/2013/845051

Abdul Rahman, H 2018. Climate change scenarios in Malaysia: Engaging the public. Int. J. Malay-Nusantara Stud., 1, 55–77.

Ablain M, B Meyssignac, L Zawadzki et al. 2019. Uncertainty in satellite estimate of global mean sea level changes, trend and acceleration. Earth Syst. Sci. Data, 11, 1189–1202. https://doi.org/10.5194/essd-11-1189-2019

Abraha-Kahsay, G, LG Hansen 2016. The effect of climate change and adaptation policy on agricultural production in Eastern Africa. Ecol. Econ., 121, 54–64.

Abrahamsen, P, S Hansen 2000. Daisy: An open soil-crop-atmosphere system model. Environ. Model. Softw., 15, 313–330. https://doi.org/10.1016/S1364-8152(00)00003-7

Ab-Rahim, R, T Xin-Di 2016. The determinants of CO_2 emissions in ASEAN+3 countries. J. Entrepreneurship Bus., 4(1), 38–49.

Acevedo, MF, DR Harvey, FG Palis 2018. Food security and the environment: Interdisciplinary research to increase productivity while exercising environmental conservation. Glob. Food Secur., 16, 127–132.

ACIA 2005. Arctic Climate Impact Assessment. ACIA Overview Report. Cambridge: Cambridge University Press, p. 1020.

Acosta, LA, DB Magcale-Macandog, KSK Kumar et al. 2016. The role of bioenergy in enhancing energy, food and ecosystem sustainability based on societal perceptions and preferences in Asia. Agriculture, 6, 19. https://doi.org/10.3390/agriculture6020019

ADB (Asian Development Bank) 2010. Methodology for Estimating Carbon Footprint of Road Projects – Case Study: India.

Adekanmbi, T, X Wang, S Basheer et al. 2023. Assessing future climate change impacts on potato yields — A case study for Prince Edward Island, Canada. Foods, 12, 1176. https://doi.org/10.3390/foods12061176

Adetunji, AT, B Ncube, R Mulidzi et al. 2020. Management impact and benefit of cover crops on soil quality: A review. Soil Tillage Res., 204, 104717. https://doi.org/10.1016/j.still.2020.104717

Adhikari, U, AP Nejadhashemi, SA Woznicki 2015. Climate change and eastern Africa: A review of impact on major crops. Food Energy Secur., 4, 110–132.

AGNES 2020. Land Degradation and Climate Change in Africa. Africa Group of Negotiators Experts Support. Policy Brief No. 2, March, 2020. https://agnesafrica.org/download/policy-brief-no-2-land-degradation-and-climate-change-in-africa/

Agnolucci, P, V De Lipsis 2020. Long-run trend in agricultural yield and climatic factors in Europe. Clim. Change, 159, 385–405. https://doi.org/10.1007/s10584-019-02622-3

AGRA 2019. Irrigation doubles African food production. Available online at: https://agra.org/irrigation-doubles-african-food-production/

Agriculture and Agri-Food Canada 2021. Potato Market Information Review 2020–2021. Available online at: https://agriculture.canada.ca/sites/default/files/documents/2022-02/potato_market_review_revue_marche_pomme_terre_2020-eng.pdf

Agronomix Software I 2022. Plant breeding software from agronomix. Available at: https://www.agronomix.com

Ahmad, B, S Mahmood 2017. Observed, Simulated and Projected Extreme Climate Indices Over Pakistan. Hamburg, Germany: Anchor Academic Publishing.

Ahmad, I, B Ahmad, K Boote et al. 2020. Adaptation strategies for maize production under climate change for semi-arid environments. Eur. J. Agron., 115, 126040.

Ahmed, I, A Ullah, MH Rahman et al. 2019. Climate Change Impacts and Adaptation Strategies for Agronomic Crops [Internet]. Climate Change and Agriculture. IntechOpen; Available from: https://doi.org/10.5772/intechopen.82697.

Ahmed, M 2022. Global Agricultural Production: Resilience to Climate Change. Cham, Switzerland: Springer International Publishing. ISBN 978-3-031-14973-3.

Aitekeyeva, N, X Li, H Guo et al. 2020. Drought risk assessment in cultivated areas of Central Asia using MODIS time-series data. Water, 12, 1738. https://doi.org/10.3390/w12061738

Aizebeokhai, AP 2009. Global warming and climate change: Realities, uncertainties and measures. Int. J. Phys. Sci., 4(13), 868–879.

Ajibola FO, B Zhou, S Shahid et al. 2022. Performance of CMIP6 HighResMIP simulations on West African drought. Front. Earth Sci., 10, 925358. https://doi.org/10.3389/feart.2022.925358

Akinnagbe, OM, I Irohibe 2014. Agriculture adaptation strategies to climate change impacts in Africa: A review. Bangladesh J. Agric. Res., 39(3), 407–418.

Akiyama, H, XY Yan, K Yagi 2010. Evaluation of effectiveness of enhanced efficiency fertilizers as mitigation options for N_2O and NO emissions from agricultural soils: Meta-analysis. Global Change Biol., 16, 1837–1846. https://doi.org/10.1111/j.1365-2486.2009.02031.x

Akter, M, H Deroo, AM Kamal et al. 2018. Impact of irrigation management on paddy soil N supply and depth distribution of abiotic drivers. Agric. Ecosyst. Environ., 261, 12–24.

Alam, M, C Siwar, A Jaafar et al. 2013. Agricultural vulnerability and adaptation to climatic changes in Malaysia: Review on paddy sector. Curr. World Environ., 8, 1–12.

Alboghdady, M, SE El-Hendawy 2016. Economic impacts of climate change and variability on agricultural production in the Middle East and North Africa region. Int. J. Clim. Change Strateg. Manag., 8(3), 463–472.

Albuquerque, FDB, MA Maraqa, R Chowdhury et al. 2020. Greenhouse gas emissions associated with road transport projects: Current status, benchmarking, and assessment tools. Transp. Res. Procedia, 48, 2018–2030.

Alexander, K, J West 2011. Water-resource efficiency in Asia and the Pacific (2011). Illawarra Health Med. Res. Inst., 185. https://ro.uow.edu.au/ihmri/185

Alexandratos, N, J Bruinsma 2012. World Agriculture Towards 2030/2050: the 2012 Revision. ESA Working Paper No. 12–03. Rome (Italy): FAO. Available at http://www.fao.org/3/a-ap106e.pdf

Al-Faraj, FAM, D Tigkas, M Scholz 2016. Irrigation efficiency improvement for sustainable agriculture in changing climate: A transboundary watershed between Iraq and Iran. Environ. Process, 3, 603–616.

Alfieri, L, B Bisselink, F Dottori et al. 2017. Global projections of river flood risk in a warmer world. Earths Future, 5, 171–182.

Ali, S, Y Liu, M Ishaq et al. 2017. Climate change and its impact on the yield of major food crops: Evidence from Pakistan. Foods, 6(6), 39. https://doi.org/10.3390/foods6060039

Ali W, MZ Hashmi, A Jamil et al. 2022. Mid-century change analysis of temperature and precipitation maxima in the Swat River Basin, Pakistan. Front. Environ. Sci., 10. https://doi.org/10.3389/fenvs.2022.973759

Alizadeh, O, Z Lin 2021. Rapid Arctic warming and its link to the waviness and strength of the westerly jet stream over West Asia. Glob. Planet. Change, 199, 103447.

Almaraz, JJ, F Mabood, X Zhou et al. 2009. Performance of agricultural systems under contrasting growing season conditions in southwestern Quebec. J. Agron. Crop Sci., 195, 319–327.

Almazroui, M, F Saeed, S Saeed et al. 2020a Projected change in temperature and precipitation over Africa from CMIP6. Earth Syst. Environ., 4, 455–475. https://doi.org/10.1007/s41748-020-00161-x

Almazroui, M, M Ashfaq, MN Islam et al. 2021a. Assessment of CMIP6 performance and projected temperature and precipitation changes over South America. Earth Syst. Environ., 5, 155–183. https://doi.org/10.1007/s41748-021-00233-6

Almazroui, M, MN Islam, F Saeed et al. 2021b. Projected changes in temperature and precipitation over the United States, Central America, and the Caribbean in CMIP6 GCMs. Earth Syst. Environ., 5(1), 1–24.

Almazroui, M, S Saeed, F Saeed et al. 2020b Projections of precipitation and temperature over the South Asian countries in CMIP6. Earth Syst. Environ., 4, 297–320. https://doi.org/10.1007/s41748-020-00157-7

Aloysius, NR, J Sheffield, JE Saiers et al. 2016. Evaluation of historical and future simulations of precipitation and temperature in central Africa from CMIP5 climate models. J. Geophys. Res. Atmos, 121, 130–152.

Alston, JM, PG Pardey 2014. Agriculture in the global economy. J. Econ. Perspect., 28(1), 121–146.

Altchenko, Y, KG Villholth 2015. Mapping irrigation potential from renewable groundwater in Africa – A quantitative hydrological approach. Hydrol. Earth Syst. Sci., 19, 1055–1067.

Altieri, MA, CI Nicholls, A Henao et al. 2015. Agroecology and the design of climate change-resilient farming systems. Agron. Sustain. Dev., 35, 869–890. https://doi.org/10.1007/s13593-015-0285-2

Al-Zu'bi, M, SW Dejene, J Hounkpe et al. 2022. African perspectives on climate change research. Nat. Clim. Change, 12(12), 1078–1084. https://doi.org/10.1038/s41558-022-01519-x

Amare, A, B Simane 2018. Does adaptation to climate change and variability provide household food security? Evidence from Muger sub-basin of the upper Blue-Nile Ethiopia. Ecol. Process., 7, 13. https://doi.org/10.1186/s13717-018-0124-x

Amarnath, G, N Alahacoon, V Smakhtin et al. 2017. Mapping Multiple Climate-Related Hazards in South Asia. Colombo, Sri Lanka. International Water Management Institute (IWMI). 41p. (IWMI Research Report 170).

Ameztegui, A, KA Solarik, JR Parkins et al. 2018. Perceptions of climate change across the Canadian forest sector: The key factors of institutional and geographical environment. PLOS One, 13(6), e0197689. https://doi.org/10.1371/journal.pone.0197689

Amoah A, E Kwablah, K Korle et al. 2020. Renewable energy consumption in Africa: The role of economic well-being and economic freedom. Energ. Sustain. Soc., 10, 32. https://doi.org/10.1186/s13705-020-00264-3

An, Z, EW Bork, X Duan et al. 2022. Quantifying past, current, and future forest carbon stocks within agroforestry systems in central Alberta, Canada. GCB Bioenergy, 14, 669–680. https://doi.org/10.1111/gcbb.12934

Anbumozhi, V, M Breiling, S Pathmarajah et al. 2012. Climate Change in Asia and the Pacific: How can Countries Adapt? New Delhi, India: Sage Publications. 400 p.

Angelopoulou, GI, CJ Koroneos, M Loizidou 2009. Environmental impacts from the construction and maintenance of a motorway in Greece, 1st International Exergy, Life Cycle Assessment, and Sustainability Workshop/Symposium (ELCAS).

Anjum, SA, W Long-chang, X Lan-lan et al. 2010. Desertification in Pakistan: Causes, impacts and management. J. Food Agric. Environ., 8(2), 1203–1208.

Arai, S, H Matsuda 2018. Key Strategies for Policymakers. In: Takeuchi K, H Shiroyama, O Saito et al. (eds) Biofuels and Sustainability. Science for Sustainable Societies. Tokyo: Springer. https://doi.org/10.1007/978-4-431-54895-9_13

Archer, ERM, WA Landman, J Malherbe et al. 2021. Managing climate risk in livestock production in South Africa: How might improved tailored forecasting contribute? Clim. Risk Manag., 32, 100312.

Argento, F, T Anken, F Abt et al. 2021. Site-specific nitrogen management in winter wheat supported by low-altitude remote sensing and soil data. Precision Agric., 22, 364–386. https://doi.org/10.1007/s11119-020-09733-3

Arias LA, F Berli, A Fontana et al. 2022. Climate change effects on grapevine physiology and biochemistry: Benefits and challenges of high altitude as an adaptation strategy. Front. Plant Sci., 13, 835425. https://doi.org/10.3389/fpls.2022.835425

Arora, NK 2019. Impact of climate change on agriculture production and its sustainable solutions. Environ. Sustain., 2, 95–96. https://doi.org/10.1007/s42398-019-00078-w

Aryal, JP 2022. Contribution of Agriculture to Climate Change and Low-Emission Agricultural Development in Asia and the Pacific. ADBI Working Paper 1340. Tokyo: Asian Development Bank Institute. https://doi.org/10.56506/WDBC4659

Aryal, JP, MB Mehrotra, ML Jat et al. 2015. Impacts of laser land leveling in rice–wheat systems of the north–western Indo-Gangetic plains of India. Food Secur., 7, 725–738.

Aryal, JP, TB Sapkota, DB Rahut et al. 2021. Climate risks and adaptation strategies of farmers in East Africa and South Asia. Sci. Rep., 11, 10489. https://doi.org/10.1038/s41598-021-89391-1

Aryal, JP, TB Sapkota, R Khurana et al. 2020. Climate change and agriculture in South Asia: Adaptation options in smallholder production systems. Environ. Dev. Sustain., 22, 5045–5075. https://doi.org/10.1007/s10668-019-00414-4

Asante, BO, RA Villano, IW Patrick et al. 2018. Determinants of farm diversification in integrated crop-livestock farming systems in Ghana. Renew. Agric. Food Syst., 33(2), 131–149.

Ashe, P, H Shaterian, L Akhov et al. 2017. Contrasting root and photosynthesis traits in a large-acreage Canadian durum variety and its distant parent of Algerian origin for assembling drought/heat tolerance attributes. Front. Chem., 5(121), 1–10.

Asong ZE, HS Wheater, B Bonsal et al. 2018. Historical drought patterns over Canada and their teleconnections with large-scale climate signals. Hydrol. Earth Syst. Sci., 22, 3105–3124. https://doi.org/10.5194/hess-22-3105-2018

Asong, ZE, M Elshamy, D Princz et al. 2019. Regional scenarios of change over Canada: Future climate projections, Hydrol. Earth Syst. Sci. Discuss. [preprint], https://doi.org/10.5194/hess-2019-249

ATPS (African Technology Policy Studies Network) 2013. Agricultural Innovations and Adaptations to Climate Change Effects and Food Security in Central Africa: Case of Cameroon, Equatorial Guinea and Central Africa Republic. ATPS WORKING PAPER Nᵒ. 79.

Atreya, K, BP Subedi, PL Ghimire et al. 2021. Agroforestry for mountain development: Prospects, challenges and ways forward in Nepal. Arch. Agric. Environ. Sci., 6, 87–99.

Auffhammer, M, V Ramanathan, JR Vincent 2012. Climate change, the monsoon, and rice yield in India. Clim. Change, 111, 411–424.

Austgulen, MH, SE Skuland, A Schjøll et al. 2018. Consumer readiness to reduce meat consumption for the purpose of environmental sustainability: Insights from Norway. Sustainability, 10, 3058. https://doi.org/10.3390/su10093058.

Awange, JL, L Ogalo, KH Bae et al. 2008. Falling Lake Victoria water levels: Is climate a contributing factor? Clim. Change, 89, 281–297.

Ayanlade, A, IA Oluwatimilehin, AA Oladimeji et al. 2021. Climate Change Adaptation Options in Farming Communities of Selected Nigerian Ecological Zones. In: Oguge, N., Ayal, D., Adeleke, L., da Silva, I. (eds) African Handbook of Climate Change Adaptation. Cham: Springer. https://doi.org/10.1007/978-3-030-45106-6_156

Ayar, PV, M Vrac, S Bastin et al. 2016. Intercomparison of statistical and dynamical downscaling models under the EURO-and MED-CORDEX initiative framework: Present climate evaluations. Clim. Dynam., 46, 1301–1329.

Ayompe, LM, SJ Davis, BN Egoh 2020. Trends and drivers of African fossil fuel CO_2 emissions 1990–2017. Environ. Res. Lett., 15, https://doi.org/10.1088/1748-9326/abc64f

Ayugi B, ZW Shilenje, H Babaousmail et al. 2022. Projected changes in meteorological drought over East Africa inferred from bias-adjusted CMIP6 models. Nat. Hazards, 113, 1151–1176. https://doi.org/10.1007/s11069-022-05341-8

Azhgaliyeva, D, DB Rahut 2022. Climate Change Mitigation: Policies and Lessons for Asia. ADBI Series on Asian and Pacific Sustainable Development, Tokyo, Japan: Asian Development Bank Institute. https://doi.org/10.56506/OJYG4210

Baba, AR, AM Abdulai 2021. Determinants of crop diversification and its effects on household food security in Northern Ghana. Arthaniti: J. Econ. Theory Pract., 20(2), 227–245. https://doi.org/10.1177/0976747920936818.

Bai, D, L Ye, Z Yang et al. 2022. Impact of climate change on agricultural productivity: A combination of spatial Durbin model and entropy approaches. Int. J. Clim. Change Strateg. Manag. https://doi.org/10.1108/IJCCSM-02-2022-0016

Baig, MB, SA Shahid, GS Straquadine 2013. Making rainfed agriculture sustainable through environmental friendly technologies in Pakistan: A review. Int. Soil Water Conserv. Res., 1(2), 36–52.

Bairagi S, H Bhandari, SK Das et al. 2021. Flood-tolerant rice improves climate resilience, profitability, and household consumption in Bangladesh. Food Policy, 105, 102183. https://doi.org/10.1016/j.foodpol.2021.102183.

Baker, E, S Saha 2018. Forest farming in Georgia, United States: Three potential crops. Ann. Agrar. Sci., 16 (3), 304–308. https://doi.org/10.1016/j.aasci.2018.04.003

Balasubramanian R, V Saravanakumar 2022. Climate Sensitivity of Groundwater Systems in South India: Does It Matter for Agricultural Income? In: Haque AKE, P Mukhopadhyay, M Nepal et al. (eds) Climate Change and Community Resilience. Springer, Singapore. https://doi.org/10.1007/978-981-16-0680-9_10.

Balehegn, M, E Kebreab, A Tolera et al. 2021. Livestock sustainability research in Africa with a focus on the environment. Anim. Front., 11, 47–56.

Bang, G 2021. The United States: Conditions for accelerating decarbonization in a politically divided country. Int. Environ. Agreements, 21, 43–58. https://doi.org/10.1007/s10784-021-09530-x

Bao Y, G Hoogenboom, R McClendon et al. 2015. Soybean production in 2025 and 2050 in the southeastern USA based on the SimCLIM and the CSM-CROPGRO-Soybean models. Clim. Res., 63(1), 73–89. https://www.jstor.org/stable/24897532

Barange M, G Merino, JL Blanchard et al. 2014. Impacts of climate change on marine ecosystem production in societies dependent on fisheries. Nat. Clim. Change, 4, 211–216. https://doi.org/10.1038/nclimate2119

Barriopedro, D, EM Fischer, J Luterbacher et al. 2011. The hot summer of 2010: Redrawing the temperature record map of Europe. Science, 332, 220–224.

Basso, B, RA Martinez-Feria, L Rill et al. 2021. Contrasting long-term temperature trends reveal minor changes in projected potential evapotranspiration in the US Midwest. Nat. Commun., 12, 1476. https://doi.org/10.1038/s41467-021-21763-7

Bayissa, Y, A Semu, X Yunqing et al. 2015. Spatio-temporal assessment of meteorological drought under the influence of varying record length: The case of Upper Blue Nile Basin, Ethiopia. Hydrol. Sci. J., 60, 1927–1942. https://doi.org/10.1080/02626667.2015.1032291

Bebeley JF, AY Kamara, JM Jibrin et al. 2022. Evaluation and application of the CROPGRO-soybean model for determining optimum sowing windows of soybean in the Nigeria savannas. Sci. Rep., 12(1), 6747. https://doi.org/10.1038/s41598-022-10505-4

Beillouin, D, B Schauberger, A Bastos et al. 2020. Impact of extreme weather conditions on European crop production in 2018. Philos. Trans. R. Soc. B, 375(1810), 20190510.

Beillouin, D, T Ben-Ari, E Malézieux et al. 2021. Positive but variable effects of crop diversification on biodiversity and ecosystem services. Glob. Change Biol., 27, 4697–4710. https://doi.org/10.1111/gcb.15747

Bekchanov, M, A Bhaduri, C Ringler 2015. Potential gains from water rights trading in the Aral Sea Basin. Agric. Water Manag., 152, 41–56.

Bell, SM, C Barriocanal, C Terrer et al. 2020. Management opportunities for soil carbon sequestration following agricultural land abandonment. Environ. Sci. Policy, 108, 104–111.

Beltran-Peña, A, P D'Odorico 2022. Future food security in Africa under climate change. Earth's Future, 10, e2022EF002651. https://doi.org/10.1029/2022EF002651

Ben-Asher, J, A Garcia y Garcia, G Hoogenboom 2008. Effect of high temperature on photosynthesis and transpiration of sweet corn (*Zea mays* L. var. rugosa). Photosynthetica, 46(4), 595–603.

Berbel J, C Gutiérrez-Martín, A Expósito 2018. Impacts of irrigation efficiency improvement on water use, water consumption and response to water price at field level. Agric. Water Manag., 203, 423–429. https://doi.org/10.1016/j.agwat.2018.02.026

Berhanu, Y, L Olav, A Nurfeta et al. 2019. Methane emissions from ruminant livestock in Ethiopia: Promising forage species to reduce CH_4 emissions. Agriculture, 9, 130. https://doi.org/10.3390/agriculture9060130

Bevacqua E, G Zappa, F Lehner et al. 2022. Precipitation trends determine future occurrences of compound hot–dry events. Nat. Clim. Change, 12, 350–355. https://doi.org/10.1038/s41558-022-01309-5

Bhatt D, S Maskey, MS Babel et al. 2014. Climate trends and impacts on crop production in the Koshi River basin of Nepal. Reg. Environ. Change, 14, 1291–1301. https://doi.org/10.1007/s10113-013-0576-6

Bhattacharyya, P, H Pathak, S Pal 2020. Impact of Climate Change on Agriculture: Evidence and Predictions. In: Climate Smart Agriculture. Green Energy and Technology. Springer, Singapore. https://doi.org/10.1007/978-981-15-9132-7_2

Bhuyan, M, M Islam, M Bhuiyan 2018. A trend analysis of temperature and rainfall to predict climate change for northwestern region of Bangladesh. Am. J. Clim. Change, 7, 115–134.

Biemans, E 2018. Water, climate and food production. Scientific justification of the information produced for the chapter 'Water and food production' of 'The Geography of Future Water Challenges' (2018), The Hague: PBL Netherlands Environmental Assessment Agency. Pbl.nl/future-water-challenges.

Bijl, DL, PW Bogaart, S Dekker et al. 2017. A physically-based model of long-term food demand. Glob. Environ. Change, 45, 47–62.

Bilir TE, M Chatterjee, KL Ebi, et al. 2014. Climate Change: Impacts, Adaptation, and Vulnerability. Part B: Regional Aspects. Contribution of Working Group II to the Fifth Assessment Report of the Intergovernmental Panel on Climate Change, Cambridge University Press, Cambridge, 1199–1265.

Bindi, M, JE Olesen 2011. The responses of agriculture in Europe to climate change. Reg. Environ. Change, 11 (Suppl. 1), 151–158. https://doi.org/10.1007/s10113-010-0173-x

Bingham, IJ, AJ Karley, PJ White et al. 2012. Analysis of improvements in nitrogen use efficiency associated with 75 years of spring barley breeding. Eur. J. Agron., 42, 49–58.

Birhanu, BZ, K Sanogo, SS Traore et al. 2023. Solar-based irrigation systems as a game changer to improve agricultural practices in sub-Sahara Africa: A case study from Mali. Front. Sustain. Food Syst., 7, 1085335. https://doi.org/10.3389/fsufs.2023.1085335

Bisaro, A, M de Bel, J Hinkel et al. 2020. Multilevel governance of coastal flood risk reduction: A public finance perspective. Environ. Sci. Policy, 112, 203–212. https://doi.org/10.1016/j.envsci.2020.05.018

Bjornlund, V, H Bjornlund, A van Rooyen 2020. Why agricultural production in Sub-Saharan Africa remains low compared to the rest of the world – A historical perspective. Int. J. Water Resour. Dev., 36(Suppl. 1), S20–S53. https://doi.org/10.1080/07900627.2020.1739512

Blair, D, A Little, M Wiens 2016. Review of Climate Change Projections for Southern Manitoba and Potential Impacts for Agriculture. University of Winnipeg, Winnipeg, MB. Available from https://www.gov.mb.ca/agriculture/environment/climate-change/pubs/climate-change-projections-and-impacts.pdf.

Blanc, E, J Reilly 2015. Climate change impacts on US crops. Choices, 30(2), 1–4.

Blanco-Canqui, HB, TM Shaver, JL Lindquist et al. 2015. Cover crops and ecosystem services: Insights from studies in temperate soils. Agron. J., 107, 2449–2474. https://doi.org/10.2134/agronj15.0086

Blignaut, JN, MR Chitiga-Mabugu, RM Mabugu 2005. Constructing a greenhouse gas emissions inventory using energy balances: The case of South Africa for 1998. J. Energy South Afr., 16(3), 21–32.

Blöschl, G, J Hall, A Viglione et al. 2019. Changing climate both increases and decreases European river floods. Nature, 573, 108–111. https://doi.org/10.1038/s41586-019-1495-6

Boardman, J, J Poesen 2006. Soil Erosion in Europe: Major Processes, Causes and Consequences. Soil Erosion in Europe John Wiley & Sons, pp. 477–487. ISBN-13 978 0-470-85910-0.

Bodirsky, BL, S Rolinski, A Biewald et al. 2015. Global food demand scenarios for the 21st century. PLOS One, 10(11), e0139201.

Bogale, GA, ZB Erena 2022. Drought vulnerability and impacts of climate change on livestock production and productivity in different agro-ecological zones of Ethiopia. J. Appl. Anim. Res., 50, 471–489.

Boko, M, I Niang, A Nyong et al. 2007. Africa. In: Parry ML, OF Canziani, JP Palutikof, (Eds.) Climate Change 2007: Impacts, Adaptation and Vulnerability, Contribution of Working Group II to the Fourth Assessment Report of the Intergovernmental Panel on Climate Change, Cambridge University Press, Cambridge, UK, pp. 433–467.

Bonsal, BR, EE Wheaton, A Meinert et al. 2011. Characterizing the surface features of the 1999–2005 Canadian prairie drought in relation to previous severe twentieth century events. Atmos. Ocean., 49, 320–338. https://doi.org/10.1080/07055900.2011.594024

Bonti-Ankomah, S, A Stamplecoskie, O Carrier-Leclerc 2017. An overview of the Canadian agriculture and agri-food system. https://publications.gc.ca/collections/collection_2018/aac-aafc/A38-1-1-2017-eng.pdf

Boone, RB, RT Conant, J Sircely et al. 2018. Climate change impacts on selected global rangeland ecosystem services. Glob. Change Biol., 24, 1382–1393.

Borchard, N, M Schirrmann, ML Cayuela et al. 2019. Biochar, soil and land-use interactions that reduce nitrate leaching and N_2O emissions: A meta-analysis. Sci. Total Environ., 651, 2354–2364.

Borken, J, W Knörr, H Helms et al. 2003. EcoTransIT: Ecological Transport Information Tool Environmental Methodology and Data.

Borrelli, P, DA Robinson, LR Fleischer et al. 2017. An assessment of the global impact of 21st century land use change on soil erosion. Nat. Commun., 8, 2013. https://doi.org/10.1038/s41467-017-02142-7

Borrelli, P, DA Robinson, P Panagos et al. 2020. Land use and climate change impacts on global soil erosion by water (2015–2070). Proc. Natl. Acad. Sci. U. S. A., 117(36), 21994–22001. https://doi.org/10.1073/pnas.2001403117

Bouchene, L, Z Cassim, H Engel et al. 2021. Green Africa: A Growth and Resilience Agenda for the Continent. McKinsey & Company: Bucharest, Romania.

Bouramdane, AA 2023. Assessment of CMIP6 multi-model projections worldwide: Which regions are getting warmer and are going through a drought in Africa and Morocco? What changes from CMIP5 to CMIP6? Sustainability, 15, 690. https://doi.org/10.3390/su15010690

Bourgeois, B, A Charles, LL Van Eerd et al. 2022. Interactive effects between cover crop management and the environment modulate benefits to cash crop yields: A meta-analysis. Can. J. Plant Sci. https://doi.org/10.1139/CJPS-2021-0177

Bowles, TM, M Mooshammer, Y Socolar et al. 2020. Long-term evidence shows that crop-rotation diversification increases agricultural resilience to adverse growing conditions in North America. One Earth, 2, 284–293.

Brás TA, J Seixas, N Carvalhais et al. 2021. Severity of drought and heatwave crop losses tripled over the last five decades in Europe. Environ. Res. Lett., 16, 065012. https://doi.org/10.1088/1748-9326/abf004

Brecht, H, S Dasgupta, B Laplante 2012. Sea level rise and storm surges: High stakes for a small number of developing countries. J. Environ. Dev., 21(1), 120–138. https://doi.org/10.1177/1070496511433601

Brilli, L, L Bechini, M Bindi et al. 2017. Review and analysis of strengths and weaknesses of agro-ecosystem models for simulating C and N fluxes. Sci. Total Environ., 598, 445–470.

Brink, C, L Hordijk, EC Van Ierland et al. 2000. Cost-effective N_2O, CH_4, and NH_3 abatement in European agriculture: interrelations between global warming and acidification policies. Paper presented at Ancillary Benefits and Costs of Greenhouse Gas Mitigation. Proc. IPCC Co-sponsored Workshop, 27–29 March 2000, Washington, D.C.

Brisson, N, C Gary, E Justes et al. 2003. An overview of the crop model STICS. Eur. J. Agron. 18, 309–332. https://doi.org/10.1016/S1161-0301(02)00110-7

Brouillet, A, B Sultan 2023. Livestock exposure to future cumulated climate-related stressors in West Africa. Sci. Rep., 13, 2698. https://doi.org/10.1038/s41598-022-22544-y

Brown, J 2017. Irrigation development as an instrument for economic growth in Saskatchewan: An economic impact analysis. M.Sc. Thesis. Saskatoon: University of Saskatchewan.

Bryngelsson D, S Wirsenius, F Hedenus et al. 2016. How can the EU climate targets be met? A combined analysis of technological and demand-side changes in food and agriculture. Food Policy 59, 152–164. https://doi.org/10.1016/j.foodpol.2015.12.012

Burke, MJ, JC Stephens 2018. Political power and renewable energy futures: A critical review. Energy Res. Soc. Sci., 35, 78–93. https://doi.org/10.1016/j.erss.2017.10.018

Busch, T, CH Cho, AGF Hoepner 2023. Corporate greenhouse gas emissions' data and the urgent need for a science-led just transition: Introduction to a thematic symposium. J. Bus. Ethics, 182, 897–901. https://doi.org/10.1007/s10551-022-05288-7

Byass, P 2009. Climate change and population health in Africa: Where are the scientists? Glob. Health Action, 11, 2. https://doi.org/10.3402/gha.v2i0.2065

Cai, Y, JS Bandara, DA Newth 2016. Framework for integrated assessment of food production economics in South Asia under climate change. Environ. Model. Softw., 75, 459–497. https://doi.org/10.1016/j.envsoft.2015.10.024

Cai, Z, RP Udawatta, CJ Gantzer et al. 2019. Economic impacts of cover crops for a Missouri Wheat–Corn–Soybean rotation. Agriculture, 9, 83. https://doi.org/10.3390/agriculture9040083

Calafat, FM, T Wahl, MG Tadesse et al. 2022. Trends in Europe storm surge extremes match the rate of sea-level rise. Nature, 603, 841–845. https://doi.org/10.1038/s41586-022-04426-5

Camargo-Alvarez, H, RJR Elliott, S Olin et al. 2022. Modelling crop yield and harvest index: The role of carbon assimilation and allocation parameters. Model. Earth Syst. Environ. https://doi.org/10.1007/s40808-022-01625-x

Cammarano D, S Jamshidi, G Hoogenboom et al. 2022. Processing tomato production is expected to decrease by 2050. Due to the projected increase in temperature. Nat. Food, 3, 437–444. https://doi.org/10.1038/s43016-022-00521-y

Campbell, B 2022. Climate Change Impacts and Adaptation Options in the Agrifood System—A Summary of the Recent Intergovernmental Panel on Climate Change Sixth Assessment Report. Rome, Italy. https://www.fao.org/documents/card/en/c/cc0425en

Campbell, ID, DG Durant, KL Hunter et al. 2014. Food Production; in Canada in a Changing Climate: Sector Perspectives on Impacts and Adaptation (eds.) F.J. Warren and D.S. Lemmen; Government of Canada, Ottawa, ON, pp. 99–134.

Campbell, JE, DB Lobell, RC Genova et al. 2008. The global potential of bioenergy on abandoned agriculture lands. Environ. Sci. Technol., 42, 5791–5794.

Canadian Climate Institute 2021. Canada's net zero future: Finding our way in the global transition. Full report available at: https://climateinstitute.ca/reports/canadas-net-zero-future/

Candel, JJL, R Biesbroek 2018. Policy integration in the EU governance of global food security. Food Sec., 10, 195–209.

Capros, P, A de Vita, N Tasios et al. 2013. EU Energy, Transport and GHG emissions trends to 2050—Reference Scenario 2013. European Commission-Directorate General for Energy, Directorate-General for Climate Action, and Directorate-General for Mobility and Transport, Publications Office of the European Union, Luxembourg, 2013.

Carew R, T Meng, WJ Florkowski et al. 2017. Climate change impacts on hard red spring wheat yield and production risk: Evidence from Manitoba, Canada. Can. J. Plant Sci., 98(3), 782–795. https://cdnsciencepub.com/doi/10.1139/cjps-2017-0135

Carleton, E 2022. Climate change in Africa: What will it mean for agriculture and food security?. International Livestock Research Institute. Available online at: https://www.ilri.org/news/climate-change-africa-what-will-it-mean-agriculture-and-food-security

Carozzi, M, R Martin, K Klumpp et al. 2022. Effects of climate change in European croplands and grasslands: Productivity, greenhouse gas balance and soil carbon storage. Biogeosciences, 19, 3021–3050, https://doi.org/10.5194/bg-19-3021-2022.

Carrijo, DR, ME Lundy, BA Linquist 2017. Rice yields and water use under alternate wetting and drying irrigation: A meta-analysis. Field Crop. Res., 203, 173–180.

Carter, JG 2011. Climate change adaptation in European cities. Curr. Opin. Environ. Sustain., 3, 193–198.

Carter, MS, P Sorensen, SO Petersen et al. 2014. Effects of green manure storage and incorporation methods on nitrogen release and N_2O emissions after soil application. Biol. Fertil. Soils, 50(8), 1233–1246. https://doi.org/10.1007/s00374-014-0936-5

Carvalho, D, PS Cardoso, A Rocha 2021. Future surface temperatures over Europe according to CMIP6 climate projections: An analysis with original and bias-corrected data. Clim. Change, 167(1–2), 10. https://doi.org/10.3390/cli9090139.

Castle, SE, DC Miller, N Merten et al. 2022. Evidence for the impacts of agroforestry on ecosystem services and human well-being in high-income countries: A systematic map. Environ. Evid., 11, 10. https://doi.org/10.1186/s13750-022-00260-4

Ceglar, A, M Zampieri, A Toreti et al. 2019. Observed northward migration of agro-climate zones in Europe will further accelerate under climate change. Earth's Future, 7(9), 1088–1101. https://doi.org/10.1029/2019EF001178

Census of Agriculture 2011. Overview of the wheat sector in Manitoba https://www150.statcan.gc.ca/n1/daily-quotidien/120510/dq120510a-eng.htm

Central Software Solutions I 2022. Prism plant breeding software. Available at: http://www.teamcssi.com

Cerdà, A, I Franch-Pardo, A Novara et al. 2022. Examining the effectiveness of catch crops as a nature-based solution to mitigate surface soil and water losses as an environmental regional concern. Earth Syst. Environ., 6, 29–44. https://doi.org/10.1007/s41748-021-00284-9

Chandio, AA, Y Jiang, A Rehman et al. 2020. Determinants of demand for credit by smallholder farmers': A farm level analysis based on survey in Sindh, Pakistan. J. Asian Bus. Econ. Stud., 28, 225–240.

Chapagain, R, TA Remenyi, RMB Harris et al. 2022. Decomposing crop model uncertainty: A systematic review. Field Crop. Res., 279, 108448.

Chapman, S, CE Birch, E Pope et al. 2020. Impact of climate change on crop suitability in sub-Saharan Africa in parameterized and convection-permitting regional climate models. Environ. Res. Lett., 15, 094086, https://doi.org/10.1088/1748-9326/ab9daf

Chapman, S, CE Birch, MV Galdos et al. 2021. Assessing the impact of climate change on soil erosion in East Africa using a convection-permitting climate model. Environ. Res. Lett., 16, 084006.

Chemura, A, L Murken, N Gloy et al. 2022. Crop diversification under climate change: A comparative assessment in Ghana, Burkina Faso, Ethiopia and Niger. Conference on International Research on Food Security, Natural Resource Management and Rural Development, Tropentag 2022 September 14–16, 2022.

Chen, F, J Wang, L Jin et al. 2009. Rapid warming in mid-latitude central Asia for the past 100 years. Front. Earth Sci. China, 3, 42–50.

Chen, L, G Msigwa, M Yang et al. 2022. Strategies to achieve A carbon neutral society: A review. Environ. Chem. Lett., 20, 2277–2310. https://doi.org/10.1007/s10311-022-01435-8

Chen, Y, GW Marek, TH Marek et al. 2021. Modeling climate change impacts on blue, green, and grey water footprints and crop yields in the Texas High Plains, USA. Agric. For. Meteorol., 310, 108649.

Chen, ZD, FS Chen 2022. Socio-economic factors influencing the adoption of low carbon technologies under rice production systems in China. Carbon Balance Manage., 17, 19. https://doi.org/10.1186/s13021-022-00218-6.

Cheng, M, B McCarl, C Fei 2022. Climate change and livestock production: A literature review. Atmosphere, 13, 140. https://doi.org/10.3390/atmos13010140.

Cheng W, L Dan, X Deng et al. 2022. Global monthly gridded atmospheric carbon dioxide concentrations under the historical and future scenarios. Sci. Data, 9, 83. https://doi.org/10.1038/s41597-022-01196-7

Chenu, C, DA Angers, P Barré et al. 2019. Increasing organic stocks in agricultural soils: Knowledge gaps and potential innovations Soil Tillage Res., 188, 41–52, https://doi.org/10.1016/j.still.2018.04.011

Chenyang, L, A Currie, H Darrin et al. 2021. Farming with trees: Reforming U.S. farm policy to expand agroforestry and mitigate climate change. Ecol. Law Q., 48, 1–40. https://doi.org/10.2139/ssrn.3717877

Chhabra, A, KR Manjunath, S Panigrahy et al. 2013. Greenhouse gas emissions from Indian livestock. Clim. Change, 117, 329–344. https://doi.org/10.1007/s10584-012-0556-8.

Chiaramonti, D, G Talluri, N Scarlat et al. 2021. The challenge of forecasting the role of biofuel in EU transport decarbonization at 2050: A meta-analysis review of published scenarios. Renew. Sustain. Energy Rev., 139, 110715.

Chizmar, S, R Parajuli, GE Frey et al. 2022. Challenges and opportunities for agroforestry practitioners to participate in state preferential property tax programs for agriculture and forestry. Trees For. People, 7, 100176, https://doi.org/10.1016/j.tfp.2021.100176

Chmielewski, FM, T Rötzer 2001. Response of tree phenology to climate change across Europe. Agric. For. Meteorol., 108, 101–112.

Choi, J, J Ko, KN An et al. 2021. Simulation of staple crop yields for determination of regional impacts of climate change: A case study in Chonnam Province, Republic of Korea. Agronomy, 11, 2544. https://doi.org/10.3390/agronomy11122544

Choi, YW, D Campbell, E Eltahir 2022. Near-term regional climate change in East Africa. Clim. Dyn., https://doi.org/10.1007/s00382-022-06591-9

Chowdhury, MMI, SM Rahman, MIUA Amran et al. 2022. Climate change impacts on food system security and sustainability in Bangladesh. Res. Sq. https://doi.org/10.21203/rs.3.rs-1673139/v1

Chowdhury, S, M Al-Zahrani 2013. Implications of climate change on water resources in Saudi Arabia. Arab. J. Sci. Eng., 38, 1959–1971. https://doi.org/10.1007/s13369-013-0565-6

Church, JA 2019. Sea-Level and Climate Change. In: Finkl CW, C Makowski (eds) Encyclopedia of Coastal Science. Encyclopedia of Earth Sciences Series. Springer, Cham. https://doi.org/10.1007/978-3-319-93806-6_382

Chynoweth, DP, JM Owens, R Legrand 2001. Renewable methane from anaerobic digestion of biomass. Renew. Energy, 22, 1–8.

Clapp, J 2017. Food self-sufficiency: Making sense of it, and when it makes sense. Food Policy, 66, 88–96. https://doi.org/10.1016/j.foodpol.2016.12.001

Clifton, C, R Evans, S Hayes et al. 2010. Water and Climate Change: Impacts on Groundwater Resources and Adaptation Options. Water Working Notes No. 25. Washington, DC, USA: World Bank. 76p.

Climate Centre 2021. Asia Pacific. Available at: https://www.climatecentre.org/wp-content/uploads/RCCC-ICRC-Country-profiles-Region_Asia_Pacific.pdf

Cole, MA, RJR Elliott, K Shimamoto 2005. Industrial characteristics, environmental regulations and air pollution: An analysis of the UK manufacturing sector. J. Environ. Econ. Manag., 50, 121–143.

Collier, P, G Conway, T Venables 2008. Climate change and Africa. Oxf. Rev. Econ. Policy, 24(2), 337–353. http://www.jstor.org/stable/23606648

Collins, JM 2011. Temperature variability over Africa. J. Clim., 24(14), 3649–3666.

Conradt, T, H Engelhardt, C Menz et al. 2023. Cross-sectoral impacts of the 2018–2019 Central European drought and climate resilience in the German part of the Elbe River basin. Reg. Environ. Change, 23, 32. https://doi.org/10.1007/s10113-023-02032-3

Cook, BI, JS Mankin, K Marvel et al. 2020. Twenty-first century drought projections in the CMIP6 forcing scenarios, Earth's Future, 8, e2019EF001461. https://doi.org/10.1029/2019ef001461

Cook, BI, TR Ault, JE Smerdon 2015. Unprecedented 21st century drought risk in the American southwest and Central plains. Sci. Adv., 1, e1400082. https://doi.org/10.1126/sciadv.1400082

Cooper, M, CD Messina 2023. Breeding crops for drought-affected environments and improved climate resilience. Plant Cell, 35(1), 162–186. https://doi.org/10.1093/plcell/koac321

Copernicus Climate Change Service 2022. Global Climate Highlights. https://climate.copernicus.eu

Cormier, F, J Foulkes, B Hirel et al. 2016. Breeding for increased nitrogen-use efficiency: A review for wheat (*T. aestivum* L.). Plant Breed., 135, 255–278. https://doi.org/10.1111/pbr.12371

Corwin, DL 2020. Climate change impacts on soil salinity in agricultural areas. Eur. J. Soil Sci., https://doi.org/10.1111/ejss.13010

Crawford, CL, H Yin, VC Radeloff 2022. Rural land abandonment is too ephemeral to provide major benefits for biodiversity and climate. Sci Adv., 8(21), eabm8999. https://doi.org/10.1126/sciadv.abm8999

Crowther, TW, KE Todd-Brown, CW Rowe et al. 2016. Quantifying global soil carbon losses in response to warming. Nature, 540(7631), 104–108.

Cruz, RV, H Harasawa, M Lal et al. 2007. Asia. Climate Change 2007: Impacts, Adaptation and Vulnerability. Contribution of Working Group II to the Fourth Assessment Report of the Intergovernmental Panel on Climate Change, M.L. Parry, O.F. Canziani, J.P. Palutikof, P.J. van der Linden and C.E. Hanson (Eds.), Cambridge University Press, Cambridge, UK, 469–506.

Curry, JA, PJ Webster 2011. Climate science and the uncertainty monster. Bull. Am. Meteorol. Soc., 92, 1667–1682.

Dadson, S, B Irvine, M Kirkby 2010. Effects of climate change on soil erosion: Estimates using newly available regional climate model data at a pan-European scale. Geophys. Res. Abstr., 12, EGU2010–EGU7047.

Daliakopoulos, I, I Tsanis, A Koutroulis et al. 2016. The threat of soil salinity: A European scale review. Sci. Total Environ., 573, 727–739.

Daloz, AS, JH Rydsaa, Ø Hodnebrog et al. 2021. Direct and indirect impacts of climate change on wheat yield in the Indo-Gangetic Plain in India. J. Agric. Food Res., 4, 100132.

Dang, HL, E Li, I Nuberg et al. 2019. Factors influencing the adaptation of farmers in response to climate change: A review. Clim. Dev., 11, 765–774. https://doi.org/10.1080/17565529.2018.1562866

Danso-Abbeam, G, G Dagunga, DS Ehiakpor et al. 2021. Crop–livestock diversification in the mixed farming systems: Implication on food security in Northern Ghana. Agric. Food Secur., 10(1), 1–14. https://doi.org/10.1186/s40066-021-00319-4

Daron, JD 2014. Regional climate messages: Southern Africa. Scientific report from the CARIAA Adaptation at Scale in Semi-Arid Regions (ASSAR) Project. Available from: https://www.weadapt.org/knowledge-base/assar

Dasgupta, S, B Laplante, C Meinsner et al. 2009. The impacts of sea level rise on developing countries: A comparative analysis. Clim. Change, 93(3–4), 379–388.

Davis, J, U Sonesson, DU Baumgartner et al. 2010. Environmental impact of four meals with different protein sources: Case studies in Spain and Sweden. Food Res. Int., 43, 1874–1884.

Davis, R, R Hirji 2019. Review of Water and Climate Change Policies in South Asia. Background Paper 2. Colombo, Sri Lanka: International Water Management Institute (IWMI). 120p. (Climate Risks and Solutions: Adaptation Frameworks for Water Resources Planning, Development and Management in South Asia). https://doi.org/10.5337/2019.203, http://www.iwmi.cgiar.org/Publications/Other/PDF/sawi-paper-2.pdf

Dayamba, SD, H Djoudi, M Zida et al. 2016. Biodiversity and carbon stocks in different land use types in the Sudanian Zone of Burkina Faso, West Africa. Agric. Ecosyst. Environ., 216, 61–72.

De Bon, H, L Temple, E Malézieux et al. 2018. Organic agriculture in Africa: A source of innovation for agricultural development. Perspective (48), 1–4. https://doi.org/10.19182/agritrop/00036

de Fraiture, C, V Smakhtin, D Bossio et al. 2007. Facing climate change by securing water for food, livelihoods and ecosystems. J. SAT Agric. Res., 4(1), 1–21.

de la Vega-Leinert, AC, RJ Nicholls, A Nasser Hassan et al. 2000. Proceedings of SURVAS Expert Workshop on: African Vulnerability and Adaptation to Impacts of Accelerated Sea-Level Rise (ASLR). National Authority on Remote Sensing and Space Sciences (NARSS), Egypt, 104 pp.

de Vries, W, L Schulte-Uebbing, H Kros et al. 2021. Spatially explicit boundaries for agricultural nitrogen inputs in the European Union to meet air and water quality targets. Sci. Total Environ., 786, 147283.

De Waegemaeker, J 2019. SalFar framework on salinization processes. A comparison of salinization processes across the North Sea Region, a report by ILVO for the Interreg III North Sea Region project Saline Farming (SalFar).

de Wasseige, C, M Tadoum, R Eba'a Atyi R 2015. The Forests of the Congo Basin – Forests and climate change. Weyrich. Belgium. 128 p.

Demenois, J, E Torquebiau, MH Arnoult et al. 2020. Barriers and strategies to boost soil carbon sequestration in agriculture. Front. Sustain. Food Syst., 4, 37.

den Elzen, MGJ, I Dafnomilis, N Forsell et al. 2022. Updated nationally determined contributions collectively raise ambition levels but need strengthening further to keep Paris goals within reach. Mitig. Adapt. Strateg. Glob. Change, 27(6), 33. https://doi.org/10.1007/s11027-022-10008-7

Deng, F, L Wang, WJ Ren et al. 2014. Enhancing nitrogen utilization and soil nitrogen balance in paddy fields by optimizing nitrogen management and using polyaspartic acid urea. Field Crop. Res., 169, 30–38.

Deng Y, T Gao, H Gao et al. 2014. Regional precipitation variability in East Asia related to climate and environmental factors during 1979–2012. Sci. Rep., 4, 5693. https://doi.org/10.1038/srep05693

Depoux, A, M Hémono, S Puig-Malet et al. 2017. Communicating climate change and health in the media. Public Health Rev., 38, 7. https://doi.org/10.1186/s40985-016-0044-1

Deressa, TT, RM Hassan, C Ringler et al. 2009. Determinants of farmers' choice of adaptation methods to climate change in the Nile Basin of Ethiopia. Glob. Environ. Change, 19(2), 248–255.

Desta, S, DL Coppock 2002. Cattle population dynamics in the southern Ethiopian rangelands, 1980–97. J. Range Manag., 55, 439–451.

Devendra, C 2011. Rainfed areas and animal agriculture in Asia: The wanting agenda for transforming productivity growth and rural poverty. Asian-Australas. J. Anim. Sci., 25(1), 122–42. https://www.animbiosci.org/journal/view.php?doi=10.5713/ajas.2011.r.09

Dhakhwa, GB, CL Campbell, S LeDuc et al. 1997. Maize growth: Assessing the effects of global warming and CO_2 fertilization with crop models. Agric. For. Meteorol., 87, 253–272.

Dhawan, V 2017. Water and Agriculture in India: Background Paper for the South Asia Expert Panel During the Global Forum for Food and Agriculture (GFFA) 2017. OAV German Asia-Pacific Business Association.

Diallo, I, MB Sylla, F Giorgi et al. 2012. Multimodel GCM-RCM ensemble-based projections of temperature and precipitation over West Africa for the early 21st century. Int. J. Geophys., 2012, 1–19. https://doi.org/10.1155/2012/972896

Didovets I, A Lobanova, V Krysanova et al. 2021. Central Asian Rivers under climate change: Impacts assessment in eight representative catchments J. Hydrol. Reg. Stud., 34 (2021), Article 100779, https://doi.org/10.1016/j.ejrh.2021.100779

Diepenbrock, C, T Tang, M Jines et al. 2022. Can we harness digital technologies and physiology to hasten genetic gain in U.S. Maize breeding? Plant Physiol., 188, 1141–1157.

Ding, Y, Z Wang, Y Sun 2008. Inter-decadal variation of the summer precipitation in East China and its association with decreasing Asian summer monsoon. Part I: Observed evidences. Int. J. Clim., 28, 1139–1161.

DMC 2009. Sri Lanka National Report on Disaster Risk, Poverty and Human Development Relationship, DMC, United Nation development Programme (UNDP) in Sri Lanka. Disaster Management Centre.

Dogan, MS 2015. Integrated Water Operations in California: Hydropower, Overdraft, and Climate Change. Masters Thesis, University of California, Davis, Davis, California, p. 98.

Dong, T, J Shang, B Qian et al. 2019. Field-scale crop seeding Date estimation from MODIS data and growing degree days in Manitoba, Canada. Remote Sens., 11, 1760. https://doi.org/10.3390/rs11151760

Dono, G, R Cortignani, D Dell'Unto et al. 2016. Winners and losers from climate change in agriculture: Insights from a case study in the Mediterranean basin. Agric. Syst., 147, 65–75. https://doi.org/10.1016/j.agsy.2016.05.013

Doocy, S, A Daniels, S Murray et al. 2013. The human impact of floods: A historical review of events 1980–2009 and systematic literature review. PLOS Curr. Disast., 5, 1–29.

Dorich, CD, RT Conant, F Albanito et al. 2020. Improving N_2O emission estimates with the global N_2O database. Curr. Opin. Environ. Sustain., 47, 13–20.

Dosio, A, L Mentaschi, EM Fischer et al. 2018. Extreme heat waves under 1.5°C and 2°C global warming. Environ. Res. Lett., 13, 054006.

Doukkali, MR, G Tharcisse, S Tudal 2018. Is land degradation neutrality possible in Africa? Policy Center for the New South.

Du, P, M Xu, R Li 2021. Impacts of climate change on water resources in the major countries along the Belt and Road. PeerJ, 9, e12201. https://doi.org/10.7717/peerj.12201

Dubois, D 2007. Uncertainty theories: a unified view IEEE Cybernetic Systems Conf. (Dublin, Ireland) Invited Paper 4–9.

Duffy, C, GG Toth, RPO Hagan et al. 2021. Agroforestry contributions to smallholder farmer food security in Indonesia. Agroforest. Syst., 95, 1109–1124. https://doi.org/10.1007/s10457-021-00632-8.

Duffy PB, P Brando, GP Asner et al. 2015. Projections of future meteorological drought and wet periods in the Amazon. Proc. Natl. Acad. Sci. U. S. A., 112, 13172–13177. https://doi.org/10.1073/pnas.1421010112

Dyson, M 2017. Sharpening focus on a global low-carbon future. Joule, 1(1), 15–17. https://doi.org/10.1016/j.joule.2017.08.003

Easterling, DR, KE Kunkel, JR Arnold et al. 2017. Precipitation change in the United States. In: Climate Science Special Report: Fourth National Climate Assessment, Volume I [Wuebbles, D.J., D.W. Fahey, K.A. Hibbard, D.J. Dokken, B.C. Stewart, and T.K. Maycock (eds.)]. U.S. Global Change Research Program, Washington, DC, USA, pp. 207–230, https://doi.org/10.7930/J0H993CC

EC 2021. EU Agricultural Outlook for Markets, Income and Environment, 2021–2031. European Commission, DG Agriculture and Rural Development: Brussels.

ECCC 2018. National Inventory Report. Environment and Climate Change Canada. https://www.canada.ca/en/environment-climate-change/services/climate-change/greenhouse-gas-emissions/inventory.html

EDF 2022. How Climate Change Will Impact U.S. Corn, Soybean and Wheat Yields: A county-level analysis of climate burdens and adaptation needs in the Midwest. https://www.edf.org/climate-change-will-slow-us-crop-yield-growth-2030

Edwards, JE, KG Hirsch 2012. Adapting Sustainable Forest Management to Climate Change: Preparing for the Future. Canadian Council of Forest Ministers Climate Change Task Force. https://www.ccfm.org/wp-content/uploads/2020/08/Adapting-sustainable-forest-management-to-climate-change-preparing-for-the-future-Full-Report.pdf

EEA 2007. Annual European Community Greenhouse Gas Inventory 1990–2005 and Inventory Report 2007, Submission to the UNFCCC Secretariat, EEA Technical report No 7/2007, European Environment Agency.

EEA 2011. Greenhouse gas emissions in Europe: a retrospective trend analysis for the period 1990–2008. EEA Report No 6/2011, European Environment Agency.

EEA 2012. Climate change, impacts and vulnerability in Europe 2012. An indicator-based report, EEA Report No 12/2012, European Environment Agency, Copenhagen, Denmark.

EEA 2017a. Annual European Union Greenhouse Gas Inventory 1990–2015 and Inventory Report 2017; European Environment Agency, 2017. Available online: https://www.eea.europa.eu/publications/european-union-greenhouse-gas-inventory-2017

EEA 2017b. Climate change, impacts and vulnerability in Europe 2016. EEA Report No 1/2017; European Environment Agency.

EEA 2019. Climate change adaptation in the agriculture sector in Europe – EEA Report No 04/2019.

EEA 2020. Overall greenhouse gas emission trends and key statistics. EEA Report No 03/2020, European Environment Agency.

EEA 2021. Trends and projections in Europe 2021. EEA Report No 13/2021, European Environment Agency.

EEA 2022. Greenhouse gas emissions from agriculture in Europe. European Environment Agency.

Eekhout, J, JD Vente 2022. Global impact of climate change on soil erosion and potential for adaptation through soil conservation. Earth-Sci. Rev., 226, 10392.

El Kenawy, AM, MF McCabe, SM Vicente-Serrano et al. 2016. Changes in the frequency and severity of hydrological droughts over Ethiopia from 1960 to 2013. CIG, 42, 145–166. https://doi.org/10.18172/cig.2931

Elahi, I, U Saee, A Wadood et al. 2022. Effect of Climate Change on Wheat Productivity. Wheat. IntechOpen; Available at: https://doi.org/10.5772/intechopen.103780.

El-Fadel, M, E Bou-Zeid 1999. Transportation GHG emissions in developing countries. Transp. Res. Part Transp. Environ., 4, 251–264.

Elrys, AS, AS Elnahal, AI Abdo et al. 2022. Traditional, modern, and molecular strategies for improving the efficiency of nitrogen use in crops for sustainable agriculture: A fresh look at an old issue. J. Soil Sci. Plant Nutr., 22, 3130–3156. https://doi.org/10.1007/s42729-022-00873-1

Emediegwu, LE, A Wossink, A Hall 2022. The impacts of climate change on agriculture in Sub-Saharan Africa: A spatial panel data approach. World Dev., 158, 105967.

Ensminger, ME, JE Oldfield, WW Heinemann 1990. Feeds and Nutrition: Formerly, Feeds and Nutrition, Complete (second ed.), Ensminger Publishing Company, Clovis, CA.

Environment and Climate Change Canada 2016. Canada's mid-century long-term low-greenhouse gas development strategy. United Nations climate change. Available at: https://unfccc.int/process/the-paris-agreement/long-term-strategies

Escarcha, JF, JA Lassa, KK Zander 2018. Livestock under climate change: A systematic review of impacts and adaptation. Climate, 6, 54. https://doi.org/10.3390/cli6030054

Estrada, F, O Calderón-Bustamante, W Botzen et al. 2022. AIRCC-Clim: A user-friendly tool for generating regional probabilistic climate change scenarios and risk measures. Environ. Model. Softw., 157, 15.

European Commission 2019. Delivering the European green deal https://ec.europa.eu/info/strategy/priorities-2019-2024/european-green-deal/delivering-european-green-deal_en

European Commission 2020. Farm to Fork Strategy for a Fair, Healthy and Environmentally-Friendly Food System. (Communication) COM (2020) 381 final, 20 May 2020, 7.

European Commission 2020. Kick-starting the journey towards a climate-neutral Europe by 2050. EU Climate Action Progress Report.

EUROSTAT 2021. Annual freshwater abstraction by source and sector. https://ec.europa.eu/eurostat/databrowser/view/ENV_WAT_ABS/default/table?lang=en

Evett, SR, PD Colaizzi, FR Lamm et al. 2020. Past, present, and future of irrigation on the U.S. Great Plains. Trans. ASABE, 63(3), 703–729. https://doi.org/10.13031/trans.13620

Exenberger, A, A Pondorfer, MH Wolters 2014. Estimating the Impact of Climate Change on Agricultural Production: Accounting for Technology Heterogeneity Across Countries. Kiel Working Paper No. 1920. Kiel Institute for the World Economy: Kiel Germany.

Eyshi Rezaei, E, S Siebert, F Ewert 2017. Climate and management interaction cause diverse crop phenology trends. Agric. For. Meteorol., 233, 55–70. https://doi.org/10.1016/j.agrformet.2016.11.003

Faiz-ul Islam, S, A de Neergaard, BO Sander et al. 2020. Reducing greenhouse gas emissions and grain arsenic and lead levels without compromising yield in organically produced rice. Agric. Ecosyst. Environ., 295, 16.

Fan, D, T Liu, F Sheng et al. 2020. Nitrogen deep placement mitigates methane emissions by regulating methanogens and methanotrophs in no-tillage paddy fields. Biol. Fertil. Soils, 56, 711–727.

Fan, J, BG McConkey, BC Liang et al. 2019. Increasing crop yields and root input make Canadian farmland a large carbon sink. Geoderma, 336, 49–58.

FAO 2015. FAO Statistical Pocketbook – World Food and Agriculture 2015.

FAO 2018. The State of Agricultural Commodity Markets 2018. Agricultural trade, climate change and food security. Rome.

FAO 2020. Evaluation of "Promotion of climate-smart livestock management integrating reversion of land degradation and reduction of desertification risks in vulnerable provinces". Project Evaluation Series, 12/2020. Rome.

FAO 2021. The Share of Agri-Food Systems in Total Greenhouse Gas Emissions: Global, Regional and Country Trends 1990–2019, Food and Agriculture Organization of the United Nations, Rome, Italy, https://fenixservices.fao.org/faostat/static/documents/EM/cb7514en.pdf

FAOSTAT 2015. Crops and Livestock Products. Food and Agriculture Organization of the United Nations, Rome, Italy, https://www.fao.org/faostat/en/

FAOSTAT 2022. Suite of Food Security Indicators. Food and Agriculture Organization, FAOSTAT Statistical Database. Food and Agriculture Organization of the United Nations.

Farooq A, N Farooq, H Akbar et al. 2023. A critical review of climate change impact at a global scale on cereal crop production. Agronomy, 13, 162. https://doi.org/10.3390/agronomy13010162

Faye, B, H Webber, T Gaiser et al. 2023. Climate change impacts on European arable crop yields: Sensitivity to assumptions about rotations and residue management. Eur. J. Agron., 142, 126670.

Fellmann T, P Witzke, F Weiss et al. 2018. Major challenges of integrating agriculture into climate change mitigation policy frameworks. Mitig. Adapt. Strateg. Glob. Change, 23, 451–468. https://doi.org/10.1007/s11027-017-9743-2

Feng, R, R Yu, H Zheng et al. 2018. Spatial and temporal variations in extreme temperature in Central Asia: variations in extreme temperature in Central Asia. Int. J. Climatol., 38(April), e388–e400.

Fenta, AA, A Tsunekawa, N Haregeweyn et al. 2019. Land susceptibility to water and wind erosion risks in the East Africa region. Sci. Total Environ., 703, 1–20.

FHWA 2015. INVEST Version 1.1 Compendium. https://www.sustainablehighways.org/files/869.pdf

Filimonova, IV, AV Komarova, VM Kuzenkova et al. 2022. Emissions of CO_2 in Europe and the Asia–pacific region: Panel data model. Energy Rep., 8, 894–901.

Finney, DM, SE Eckert, JP Kaye 2015. Drivers of nitrogen dynamics in ecologically based agriculture revealed by long-term, high-frequency field measurements. Ecol. Appl., 25(8), 2210–2227. https://doi.org/10.1890/14-1357.1.sm

Fischer, EM, R Knutti 2016. Observed heavy precipitation increase confirms theory and early models. Nat. Clim. Change, 6, 986–991. https://doi.org/10.1038/nclimate3110

Fisher, AC, WM Hanemann, MJ Roberts et al. 2012. The economic impacts of climate change: Evidence from agricultural output and random fluctuations in weather: Comment. Am. Econ. Rev., 102(7), 3749–3760.

Folberth, C, J Elliott, C Müller et al. 2019. Parameterization-induced uncertainties and impacts of crop management harmonization in a global gridded crop model ensemble. PLOS One, 14(9), e0221862. https://doi.org/10.1371/journal.pone.0221862

Folger, P 2017. Drought in the United States: Causes and current understanding. Cong. Res. Serv., 7–5700.

Fonta, WM, ET Ayuk, T van Huysen 2018. Africa and the green climate fund: Current challenges and future opportunities. Clim. Policy, 18, 1210–1225. https://doi.org/10.1080/14693062.2018.1459447

Ford, T, CF Labosier 2014. Spatial patterns of drought persistence in the Southeastern United States. Int. J. Climatol., 34, 2229–2240.

Fouli, Y, M Hurlbert, R Kröbel 2021. Greenhouse gas emissions from Canadian agriculture: Estimates and measurements. The Simpson Centre, SPP Briefing Paper Volume 14, 35.

Franke JA, C Müller, J Elliott et al. 2020. The GGCMI phase 2 experiment: Global gridded crop model simulations under uniform changes in CO_2, temperature, water, and nitrogen levels (protocol version 1.0). Geosci. Model. Dev., 13, 2315–2336. https://doi.org/10.5194/gmd-13-2315-2020

Freibauer, A 2003. Regionalised inventory of biogenic greenhouse gas emissions from European agriculture. Eur. J. Agron., 19, 135–160.

Freibauer, A, M Kaltschmitt 2001. Biogenic Greenhouse Gas Emissions from Agriculture in Europe. University of Stuttgart, Institute of Energy Economics and the Rational Use of Energy, Forschungsbericht N8. 78, February 2001.

Friedli CN, S Abiven, D Fossati et al. 2019. Modern wheat semi-dwarfs root deep on demand: Response of rooting depth to drought in a set of Swiss era wheats covering 100 years of breeding. Euphytica, 215, 85. https://doi.org/10.1007/s10681-019-2404-7

Friedrich, J, T Damassa 2014. The History of Carbon Dioxide Emissions. World Resources Institute, May 21, 2014, https://www.wri.org/insights/history-carbon-dioxide-emissions

Frisvold, G, C Sanchez, N Gollehon et al. 2018. Evaluating gravity-flow irrigation with lessons from Yuma, Arizona, USA. Sustainability, 10, 1548. https://doi.org/10.3390/su10051548

Fulton, L, A Mejia, M Arioli et al. 2017. Climate change mitigation pathways for Southeast Asia: CO_2 emissions reduction policies for the energy and transport sectors. Sustainability, 9(7), 1160. https://doi.org/10.3390/su9071160

Funk C 2011. We thought trouble was coming. Nature, 476, 7. https://doi.org/10.1038/476007a

Funk, C, F Davenport, L Harrison et al. 2018. Anthropogenic enhancement of moderate-to-strong El Niño events likely contributed to drought and poor harvests in Southern Africa during 2016. Bull. Am. Meteorol. Soc., 99(1), S91–S96.

Gabriel, A, M Gandorfer 2023. Adoption of digital technologies in agriculture—an inventory in a European small-scale farming region. Precision Agric., 24, 68–91. https://doi.org/10.1007/s11119-022-09931-1

Gadédjisso-Tossou, A, T Avellán, N Schütze 2020. Impact of irrigation strategies on maize (*Zea mays* L.) Production in the Savannah region of Northern Togo (West Africa). Water SA, 46(1), 141–152. https://doi.org/10.17159/wsa/2020.v46.i1.7894

Gaetani, M, S Janicot, M Vrac et al. 2020. Robust assessment of the time of emergence of precipitation change in West Africa. Sci. Rep., 10, 7670. https://doi.org/10.1038/s41598-020-63782-2

Gaffney, J, J Schussler, C Löffler et al. 2015. Industry-scale evaluation of maize hybrids selected for increased yield in drought-stress conditions of the US corn belt. Crop Sci., 55, 1608–1618.

Gamelin BL, J Feinstein, J Wang et al. 2022. Projected U.S. drought extremes through the twenty-first century with vapor pressure deficit. Sci. Rep., 12, 8615. https://doi.org/10.1038/s41598-022-12516-7

Ganguli, P, AR Ganguly 2016. Space-time trends in U.S. meteorological droughts. J. Hydrol.: Reg. Stud., 8, 235–259.

Garbelini, LG, H Debiasi, J Alvadi et al. 2022. Diversified crop rotations increase the yield and economic efficiency of grain production systems. Eur. J. Agron., 137, 126528. https://doi.org/10.1016/J.EJA.2022.126528

García-León, D, A Casanueva, G Standardi et al. 2021. Current and projected regional economic impacts of heatwaves in Europe. Nat. Commun., 12, 1, https://doi.org/10.1038/s41467-021-26050-z.

Garnett, T 2009. Livestock-related greenhouse gas emissions: Impacts and options for policymakers. Environ. Sci. Policy, 12, 491–503.

Gaudin, ACM, TN Tolhurst, AP Ker 2015. Increasing crop diversity mitigates weather variations and improves yield stability. PLOS One, 10, e0113261.

Gautam, M, M Agrawal 2021. Greenhouse Gas Emissions from Municipal Solid Waste Management: A Review of Global Scenario. In: Muthu, S.S. (eds) Carbon Footprint

Case Studies. Environmental Footprints and Eco-design of Products and Processes. Springer, Singapore. https://doi.org/10.1007/978-981-15-9577-6_5

Ge, M, J Friedrich, L Vigna 2020. 4 Charts Explain Greenhouse Gas Emissions by Countries and Sectors. World Resources Institute. Available at: https://www.wri.org/insights/4-charts-explain-greenhouse-gas-emissions-countries-and-sectors

Gebremeskel Haile, G, Q Tang, S Sun et al. 2019. Droughts in East Africa: Causes, impacts and resilience. Earth-Sci. Rev., 193, 146–161.

Genius, M, P Koundouri, C Nauges 2014. Information transmission in irrigation technology adoption and diffusion: Social learning, extension services and spatial effects. Am. J. Agric. Econ., 96, 328–344.

Gerber, P, H Steinfeld, B Henderson et al. 2013. Tackling Climate Change Through Livestock: A Global Assessment of Emissions and Mitigation Opportunities Food and Agriculture Organization of the United Nations (FAO).

Ghosh, P, K Hazra, M Venkatesh et al. 2020. Grain legume inclusion in cereal–cereal rotation increased base crop productivity in the long run. Exp. Agric., 56(1), 142–158. https://doi.org/10.1017/S0014479719000243

Gil-Alana, LA, L Sauci 2019. US temperatures: Time trends and persistence. Int. J. Climatol., 39(13), 5091–5103. https://doi.org/10.1002/joc.6128

Gilani, HR, JL Innes 2020. The state of Canada's forests: A global comparison of the performance on Montréal process criteria and indicators. For. Policy Econ., 118, 102234. https://doi.org/10.1016/j.forpol.2020.102234

Gill, JS, PWG Sale, RR Peries et al. 2009. Changes in soil physical properties and crop root growth in dense sodic subsoil following incorporation of organic amendments. Field Crops Res., 114, 137–146.

Ginbo, T 2022. Heterogeneous impacts of climate change on crop yields across altitudes in Ethiopia. Clim. Change, 170, 12. https://doi.org/10.1007/s10584-022-03306-1

Giorgi, F 2006. Climate change hot-spots. Geophys. Res. Lett., 33, 1–4.

Giorgi, F 2010. Uncertainties in climate change projections, from the global to the regional scale. EPJ Web Conf., 9, 115–129.

Girvetz, E, J Ramirez-Villegas, L Claessens et al. 2019. Future Climate Projections in Africa: Where Are We Headed?. In: Rosenstock T, A Nowak, E Girvetz (eds) The Climate-Smart Agriculture Papers. Springer, Cham. https://doi.org/10.1007/978-3-319-92798-5_2

Gkatzoflias, D, C Kouridis, L Ntziachristos et al. 2007. COPERT 4, Computer Programme to Calculate Emissions from Road Transport. Aristotle University Thessaloniki/European Environment Agency.

Gkoumas, K, FL Marques dos Santos, M Stepniak et al. 2021. Research and innovation supporting the European sustainable and smart mobility strategy: A technology perspective from recent European Union projects. Appl. Sci., 11, 11981. https://doi.org/10.3390/app112411981

Global Commission on Adaptation 2019. Adapt now: a global call for leadership on climate resilience. https://gca.org/wp-content/uploads/2019/09/GlobalCommission_Report_FINAL.pdf

Gobinath, R, G Ganapathy, E Gayathiri et al. 2022. Ecoengineering practices for soil degradation protection of vulnerable hill slopes. In Computers in Earth and Environmental Sciences; Elsevier: Amsterdam, The Netherlands, pp. 255–270.

Godde, CM, D Mason-D'Croz, DE Mayberry et al. 2021. Impacts of climate change on the livestock food supply chain; A review of the evidence. Glob Food Sec., 28, 100488. https://doi.org/10.1016/j.gfs.2020.100488

Godde CM, R Boone, AJ Ash et al. 2020. Global rangeland production systems and livelihoods at threat under climate change and variability. Environ. Res. Lett., 15, 44021. https://doi.org/10.1088/1748-9326/ab7395

Gornall, J, R Betts, E Burke et al. 2010. Implications of climate change for agricultural productivity in the early twenty-first century. Philos. Trans. R. Soc. Lond. B Biol. Sci., 365(1554), 2973–2989. https://doi.org/10.1098/rstb.2010.0158

Gould, I, JD Waegemaeker, D Tzemi et al. 2021. Salinization threats to agriculture across the North Sea region. In Negacz, K. et al., Future of Sustainable Agriculture in Saline Environments, Taylor &Francis Group, pp. 71–92.

Govindasamy, P, SK Muthusamy, M Bagavathiannan et al. 2023. Nitrogen use efficiency— A key to enhance crop productivity under A changing climate. Front. Plant Sci., 14, 1121073. https://doi.org/10.3389/fpls.2023.1121073

Gowda, P, JL Steiner, C Olson et al. 2018. Agriculture and Rural Communities Impacts, Risks, and Adaptation in the United States: Fourth National Climate Assessment vol II, ed D R Reidmiller, C W Avery, D R Easterling, K E Kunkel, K L M Lewis, T K Maycock and B C Stewart (Washington, DC: US Global Change Research Program), pp. 391–437.

Greaves, EG, YM Wang 2016. Assessment of FAO AquaCrop model for simulating maize growth and productivity under deficit irrigation in a tropical environment. Water, 8, https://doi.org/10.3390/w8120557

Green, R, J Milner, AD Dangour et al. 2015. The potential to reduce greenhouse gas emissions in the UK through healthy and realistic dietary change Clim. Change, 129, 15–17.

Green, TR, H Kipka, O David et al. 2018. Where is the USA Corn Belt, and how is it changing? Sci. Total Environ., 618, 1613–1618. https://doi.org/10.1016/j.scitotenv.2017.09.325

Greenan, BJW, TS James, JW Loder et al. 2019. Changes in oceans surrounding Canada; Chapter 7 in (eds.) Bush and Lemmen, Canada's Changing Climate Report; Government of Canada, Ottawa, Ontario, pp. 343–423.

Gregory RD, SG Willis, F Jiguet et al. 2009. An indicator of the impact of climatic change on European bird populations. PLoS One, 4(3), e4678. https://doi.org/10.1371/journal.pone.0004678

Gross A, T Bromm, B Glaser 2021. Soil organic carbon sequestration after biochar application: A global meta-analysis. Agronomy, 11, 2474. https://doi.org/10.3390/agronomy11122474

Grunewald, K, J Scheithauer 2010. Europe's southernmost glaciers: Response and adaptation to climate change. J. Glaciol., 56(195), 129–142. https://doi.org/10.3189/002214310791190947

Gua Mpanga, IK, OJ Idowu 2021. A decade of irrigation water use trends in southwest USA: The role of irrigation technology, Best management practices, and outreach education programs. Agric. Water Manag., 243, 106438.

Guarin, JR, P Martre, F Ewert et al. 2022. Evidence for increasing global wheat yield potential. Environ. Res. Lett., 17, 124045.

Guilpart, N, T Iizumi, D Makowski 2022. Data-driven projections suggest large opportunities to improve Europe's soybean self-sufficiency under climate change. Nat. Food. https://doi.org/10.1038/s43016-022-00481-3

Guntoro, B, QN Hoang, AQ A'yun et al. 2019. Dynamic responses of livestock farmers to smart farming. IOP Conf. Ser.: Earth Environ. Sci., 372, 012042.

Gupta, K, R Kumar, KK Baruah et al. 2021. Greenhouse gas emission from rice fields: A review from Indian context. Environ. Sci. Pollut. Res., 28, 30551–30572. https://doi.org/10.1007/s11356-021-13935-1

Gupta, R, E Somanathan, S Dey 2017. Global warming and local air pollution have reduced wheat yields in India. Clim. Change, 140, 593–604.

Gupta, R, R Bhattarai, H Dokoohaki et al. 2023. Sustainability of cover cropping practice with changing climate in Illinois. J. Environ. Manag., 339, 117946. https://doi.org/10.1016/j.jenvman.2023.117946

Guyomard, H, Z Bouamra-Mechemache, V Chatellier et al. 2021. Review: Why and how to regulate animal production and consumption: The case of the European Union. Animal, 15, 100283.

Haag, I, PD Jones, C Samimi 2019. Central Asia's changing climate: How temperature and precipitation have changed across time, space, and altitude. Climate, 7(10), 123. https://doi.org/10.3390/cli7100123

Habib-ur-Rahman, M, A Ahmad, A Raza et al. 2022. Impact of climate change on agricultural production; Issues, challenges, and opportunities in Asia. Front. Plant Sci., 13, 925548. https://doi.org/10.3389/fpls.2022.925548

Haensler, A, F Saeed, D Jacob 2013. Assessing the robustness of projected precipitation changes over central Africa on the basis of a multitude of global and regional climate projections. Clim. Change, 121, 349–363.

Haider, H 2019. Climate Change in Nigeria: Impacts and Responses. K4D Helpdesk Report; Institute of Development Studies: Brighton, UK.

Haile, GG, Q Tang, SM Hosseini-Moghari et al. 2020. Projected impacts of climate change on drought patterns over East Africa. Earth's Future, 8, e2020EF001502.

Haj-Amor Z, L Dhaouadi L, DG Kim et al. 2020. Effects of climate change on key soil characteristics and strategy to enhance climate resilience of smallholder farming: An analysis of a pomegranate-field in a coastal Tunisian oasis. Environ. Earth Sci., 79, 470. https://doi.org/10.1007/s12665-020-09222-w

Haj-Amor, Z, MK Ibrahimi, N Feki et al. 2016. Soil salinization and irrigation management of date palms in a Saharan environment. Environ. Monit. Assess., 188(8), 1–17.

Haj-Amor, Z, T Araya, DG Kim et al. 2022. Soil salinity and its associated effects on soil microorganisms, greenhouse gas emissions, crop yield, biodiversity and desertification: A review. Sci. Total Environ., 843, 156946. https://doi.org/10.1016/j.scitotenv.2022.156946

Haj-Amor, Z, TK Acharjee, L Dhaouadi et al. 2020. Impacts of climate change on irrigation water requirement of date palms under future salinity trend in costal aquifer of Tunisian oasis. Agric. Water Manag., 228, 105843.

Halmemies-Beauchet-Filleau, A, M Rinne, M Lamminen et al. 2018. Review: Alternative and novel feeds for ruminants: Nutritive value, product quality and environmental aspects. Animal, 12(s2), S295–S309.

Hamed, MM, MS Nashwan, S Shahid 2022. A novel selection method of CIMP6 GCMs for robust climate projection. Int. J. Climatol. https://doi.org/10.1002/joc.7461

Hameed, F, J Xu, SF Rahim et al. 2019. Optimizing nitrogen options for improving nitrogen use efficiency of rice under different water regimes. Agronomy, 9, 39. https://doi.org/10.3390/agronomy9010039

Hamidov A, K Helming, G Bellocchi et al. 2018. Impacts of climate change adaptation options on soil functions: A review of European case-studies. Land Degrad. Dev., 29(8), 2378–2389. https://doi.org/10.1002/ldr.3006

Han, X, FY Wei 2010. The influence of vertical atmospheric circulation pattern over east Asia on summer precipitation in the east of China and its forecasting test (in Chinese). Chin. J. Atmosp. Sci., 34, 533–547.

Han, Y, K Wang, Z Liu et al. 2018. Golden seed breeding cloud platform for the management of crop breeding material and genealogical tracking. Comput. Electron. Agric., 152, 206–214. https://doi.org/10.1016/j.compag.2018.07.015

Han, Y, W Ma, B Zhou et al. 2020. Effects of straw-return method for the Maize–Rice rotation system on soil properties and crop yields. Agronomy, 10, 461. https://doi.org/10.3390/agronomy10040461

Hansen, J, M Sato, R Ruedy et al. 2000. Global warming in the twenty-first century: An alternative scenario. Proc. Natl. Acad. Sci. U. S. A., 97(18), 9875–9880.

Harrison, B, J Daron, M Palmer 2021. Future sea-level rise projections for tide gauge locations in South Asia. Environ. Res. Commun., 3, 115003.

Harter, J, HM Krause, S Schuettler et al. 2014. Linking N_2O emissions from biochar-amended soil to the structure and function of the N-cycling microbial community. ISME J., 8, 660–674, https://doi.org/10.1038/ismej.2013.160

Haruna, SI, SH Anderson, RP Udawatta et al. 2020. Improving soil physical properties through the use of cover crops: A review Agrosyst. Geosci. Environ., 3 (1), e20105, https://doi.org/10.1002/agg2.20105

Harvey, S, M Zhang, GJ Fochesatto 2021. Projections of spring wheat growth in Alaska: Opportunity and adaptations in a changing climate. Clim. Serv., 22, 100235.

Haseeb, A, Y Jonathan, AG de William et al. 2019. Economic burden of livestock disease and drought in Northern Tanzania. J. Dev. Agric. Econ., 11, 140–151.

Hasegawa T, H Wakatsuki, H Ju et al. 2022. A global dataset for the projected impacts of climate change on four major crops. Sci. Data, 9, 58 https://doi.org/10.1038/s41597-022-01150-7.

Hasukawa, H, Y Inoda, S Toritsuka et al. 2021. Effect of paddy-upland rotation system on the net greenhouse gas balance as the sum of methane and nitrous oxide emissions and soil carbon storage: A case in Western Japan. Agriculture, 11, 52. https://doi.org/10.3390/agriculture11010052

Hatfield, J, G Takle, R Grotjahn et al. 2014. Climate change impacts in the United States: The third national climate assessment. US Congress: 150–174.

Hatna, E, MM Bakker 2011. Abandonment and expansion of arable land in Europe. Ecosystems, 14, 720–731. https://doi.org/10.1007/s10021-011-9441-y

Hawes, MC, U Gunawardena, S Miyasaka et al. 2000. The role of root border cells in plant defense. Trends Plant Sci., 5, 128–133.

Hayashi, K, L Llorca, S Rustini et al. 2018. Reducing vulnerability of rainfed agriculture through seasonal climate predictions: A case study on the rainfed rice production in Southeast Asia. Agric. Syst., 162, 66–76. https://doi.org/10.1016/j.agsy.2018.01.007

Hazra, KK, MS Venkatesh, PK Ghosh et al. 2014. Long–term effect of pulse crops inclusion on soil–plant nutrient dynamics in puddled rice (*Oryza sativa* L.)–wheat (*Triticum aestivum* L.) cropping system on an Inceptisol of Indo–Gangetic plain zone of India. Nutr. Cycl. Agroecosyst., 100, 95–110.

He, DC, YL Ma, ZZ Li et al. 2021. Crop rotation enhances agricultural sustainability: From an empirical evaluation of eco-economic benefits in rice production. Agriculture, 11, 91. https://doi.org/10.3390/agriculture11020091

He, W, JY Yang, B Qian et al. 2018. Climate change impacts on crop yield, soil water balance and nitrate leaching in the semiarid and humid regions of Canada. PLOS One, 13(11), e0207370. https://doi.org/10.1371/journal.pone.0207370

He, Y, H Wang, B Qian et al. 2012. How early can the seeding dates of spring wheat be under current and future climate in Saskatchewan, Canada? PLoS One, 7, e45153, https://doi.org/10.1371/journal.pone.0045153

Hedenus, F, S Wirsenius, DJA Johansson 2014. The importance of reduced meat and dairy consumption for meeting stringent climate change targets. Clim. Change, 124, 79–91.

Hellegers, P, G van Halsema 2021. SDG indicator 6.4.1 "Change in water use efficiency over time": Methodological flaws and suggestions for improvement. Sci. Total Environ., 801, 149431.

Hemathilake, DMKS, DMCC Gunathilake 2022. Agricultural productivity and food supply to meet increased demands. In Future Foods; Academic Press: Cambridge, MA, USA, 2022; 539–553. Available online: https://www.sciencedirect.com/science/article/pii/B9780323910019000165.

Heno, S, L Viou, MFR Khan 2018. Sugar beet production in France. Sugar Tech., 20, 392–395. https://doi.org/10.1007/s12355-017-0575-x

Hens, L, NA Thinh, TH Hanh et al. 2018. Sea-level rise and resilience in Vietnam and the Asia-Pacific: A synthesis. Vietnam J. Earth Sci., 40, 126–152.

Herrero M, D Mason-D'Croz, PK Thornton et al. 2023. Livestock and Sustainable Food Systems: Status, Trends, and Priority Actions. In: von Braun, J., Afsana, K., Fresco, L.O., Hassan, M.H.A. (eds) Science and Innovations for Food Systems Transformation. Springer, Cham. https://doi.org/10.1007/978-3-031-15703-5_20

Heureux, AMC, J Alvar-Beltrán, R Manzanas et al. 2022. Climate trends and extremes in the Indus River Basin, Pakistan: Implications for agricultural production. Atmosphere, 13, 378. https://doi.org/10.3390/atmos13030378

Heyl, K, T Döring, B Garske et al. 2020. The common agricultural policy beyond 2020: A critical review in light of global environmental goals. Rev. Eur. Comp. Int. Environ. Law. https://doi.org/10.1111/reel.12351

Hinkel, J, S Brown, L Exner et al. 2012. Sea-level rise impacts on Africa and the effects of mitigation and adaptation: An application of DIVA. Reg. Environ. Change, 12, 207–224.

Holka, M, J Kowalska, M Jakubowska 2022. Reducing carbon footprint of agriculture—Can organic farming help to mitigate climate change? Agriculture, 12, 1383. https://doi.org/10.3390/agriculture12091383

Holst, R, XH Yu, G Grün 2013. Climate change, risk and grain yields in China. J. Integr. Agric., 12, 1279–1291.

Holzkämper, A, P Calanca, J Fuhrer 2012. Statistical crop models: Predicting the effects of temperature and precipitation changes. Clim. Res., 51, 11–21.

Hoogenboom, G, CH Porter, KJ Boote et al. 2019. The DSSAT crop modeling ecosystem. In: p. 173–216 [K.J. Boote, editor] Advances in Crop Modeling for a Sustainable Agriculture. Burleigh Dodds Science Publishing, Cambridge, United Kingdom (https://doi.org/10.19103/AS.2019.0061.10)

Hoorfar, J 2014. Global Safety of Fresh Produce: A Handbook of Best Practice, Innovative Commercial Solutions and Case Studies, 1st Edition. Woodhead Publishing Series in Food Science, Technology and Nutrition, ISBN 978-1-78242-018-7.

Horowitz, J, G Jessica 2010. The Role of Agriculture in Reducing Greenhouse Gas Emissions, EB-15, U.S. Department of Agriculture, Economic Research Service, September 2010.

Horton, RM, JS Mankin, C Lesk et al. 2016. A review of recent advances in research on extreme heat events. Curr. Clim. Change Rep., 2, 242–259.

Hossain A, TJ Krupnik, J Timsina 2020. Agricultural Land Degradation: Processes and Problems Undermining Future Food Security. In: Environment, Climate, Plant and Vegetation Growth. Springer, Cham. https://doi.org/10.1007/978-3-030-49732-3_2

Hossain, SS, H Delin, M Mingying 2022. Aftermath of climate change on Bangladesh economy: An analysis of the dynamic computable general equilibrium model. J. Water Clim. Change, 13(7), 2597–2609. https://doi.org/10.2166/wcc.2022.412.

Howden, SM, JF Soussana, FN Tubiello et al. 2007. Adapting agriculture to climate change. Proc. Natl. Acad. Sci. U. S. A., 104, 19691–19696.

Hristov, J, A Toreti, I Pérez Domínguez et al. 2020. Analysis of Climate Change Impacts on EU Agriculture by 2050, EUR 30078 EN, Publications Office of the European Union, Luxembourg, 2020, ISBN 978-92-76-10617-3, https://doi.org/10.2760/121115, JRC119632.

Hsiang, S, R Kopp, A Jina et al. 2017. Estimating economic damage from climate change in the United States. Science, 356 (6345), 1362–1369, https://doi.org/10.1126/science:aal4369

Hu, Y, G Barmeier, U Schmidhalter 2021. Genetic variation in grain yield and quality traits of spring malting barley. Agronomy, 11, 1177. https://doi.org/10.3390/agronomy11061177

Huérfano, X, T Fuertes-Mendizábal, MK Dunabeitia et al. 2015. Splitting the application of 3,4-dimethylpyrazole phosphate (DMPP): Influence on greenhouse gases emissions and wheat yield and quality under humid Mediterranean conditions. Eur. J. Agronomy, 64, 47–57.

Huppmann, D, J Rogelj, E Kriegler et al. 2018. A new scenario resource for integrated 1.5°C research. Nat. Clim. Change 8: 1027–1030. https://doi.org/10.1038/s41558-018-0317-4

Hurst, P, P Termine, M Karl 2005. Agricultural Workers and Their Contribution to Sustainable Agriculture and Rural Development. FAO, Rome. https://www.fao.org/3/af164e/af164e00.htm

Hussain, T, HT Gollany, N Hussain et al. 2022. Synchronizing nitrogen fertilization and planting date to improve resource use efficiency, productivity, and profitability of upland rice. Front. Plant Sci., 13, 895811. https://doi.org/10.3389/fpls.2022.895811

Hutchings NJ, P Sørensen, CMdS Cordovil et al. 2020. Measures to increase the nitrogen use efficiency of European agricultural production. Glob. Food Sec., 26, 100381, https://doi.org/10.1016/j.gfs.2020.100381

Huynh, HTL, LN Thi, ND Hoang 2020. Assessing the impact of climate change on agriculture in Quang Nam Province, Viet Nam using modeling approach. Int. J. Clim. Change Strateg. Manag., 12 (5), 757–771. https://doi.org/10.1108/IJCCSM-03-2020-0027

Ibrahim, MM, AA El-Baroudy, MA Taha 2016. Irrigation and fertigation scheduling under drip irrigation for maize cropping sandy soil. Int. Agrophys., 30, 47–55.

IEA 2019. CO_2 Emissions from Fuel Combustion. International Energy Agency- Paris. IEA 2021. Countries and regions https://www.iea.org/countries.

IEA 2021. Net Zero by 2050: A roadmap for the global energy sector.

IFAD 2021. What Can Smallholder Farmers Grow in a Warmer World? Climate Change and Future Crop Suitability in East and Southern Africa. International Fund for Agricultural Development. https://www.ifad.org/documents/38714170/42164624/utc_report.pdf/89c0228f-bef3-24f7-8d02-0289971d4607?t=1634830791584

IFPRI 2009. Climate Change Impact on Agriculture and Costs of Adaptations. Food Policy Report. International Food Policy Research Institute.

Iglesias, A, L Garrote 2015. Adaptation strategies for agricultural water management under climate change in Europe. Agric. Water Manag., 155, 113–124.

Iizumi, T, J Furuya, Z Shen et al. 2017. Responses of crop yield growth to global temperature and socioeconomic changes. Sci. Rep., 7, 7800. https://doi.org/10.1038/s41598-017-08214-4

Iizumi, T, M Yokozawa, M Nishimori 2011. Probabilistic evaluation of climate change impacts on paddy rice productivity in Japan. Clim. Change, 107, 391–415. https://doi.org/10.1007/s10584-010-9990-7

Institute of Climate Change and Sustainable Development of Tsinghua University 2022. End-Use Energy Consumption & CO_2 Emissions. In: China's Long-Term Low-Carbon Development Strategies and Pathways. Springer, Singapore. https://doi.org/10.1007/978-981-16-2524-4_3

Integrated Breeding Platform I 2019. Bms pro. Available at: https://www.integratedbreeding.net

Ionita, M, V Nagavciuc 2021. Changes in drought features at the European level over the last 120 years. Nat. Hazards Earth Syst. Sci., 21, 1685–1701.

Ionita, M, V Nagavciuc, P Scholz et al. 2022. Long-term drought intensification over Europe driven by the weakening trend of the Atlantic Meridional Overturning Circulation. J. Hydrol.: Reg. Stud., 42, 101176.

IPCC 2006. IPCC Guidelines for National Greenhouse Gas Inventories. Intergovernmental Panel on Climate Change (prepared by the National Greenhouse Gas Inventories Programme, Eggleston H.S., Buendia L., Miwa K., Ngara T. and Tanabe K. (eds). (IGES: Japan, 2006), 16.

IPCC 2007. Intergovernmental Panel on Climate Change (IPCC). In: Metz B, Davidson OR, Bosch PR, Dave R, Meyer LA, editors. Climate Change 2007: Mitigation. Contribution of Working Group III to the Fourth Assessment Report of the Intergovernmental Panel in Climate Change. Cambridge, United Kingdom and New York, NY, USA: Cambridge University Press; 2007.

IPCC 2013. Summary for Policymakers. In Climate Change 2013: The Physical Science Basis. Contribution of Working Group I to the Fifth Assessment Report of the Intergovernmental Panel on Climate Change. Cambridge, United Kingdom and New York, NY, USA: Cambridge University Press.

IPCC 2014. Climate Change 2014: Mitigation of Climate Change. Contribution of Working Group III to the Fifth Assessment Report of the Intergovernmental Panel on Climate Change [Edenhofer, O., R. Pichs-Madruga, Y. Sokona, E. Farahani, S. Kadner, K. Seyboth, A. Adler, I. Baum, S. Brunner, P. Eickemeier, B. Kriemann, J. Savolainen, S. Schlömer, C. von Stechow, T. Zwickel and J.C. Minx (eds.)]. Cambridge, United Kingdom and New York, NY, USA: Cambridge University Press.

IPCC 2018. Impacts of 1.5oC of global warming on natural and human systems. In: Glob. Warm. 1.5oC. An IPCC Spec. Rep. Impacts Glob. Warm. 1.5oC above Pre-industrial Levels Relat. Glob. Greenh. Gas Emiss. Pathways, Context Strength. Glob. Response to Threat Clim. Chang.

IPCC 2019. IPCC Special Report on the Ocean and Cryosphere in a Changing Climate. IPCC.

IPCC 2021. Climate Change 2021: the Physical Science Basis. IPCC.

IPCC 2022. Climate Change 2022 impacts, adaptation and vulnerability. Contribution of Working Group II to the Sixth Assessment Report of the Intergovernmental Panel on Climate Change. 2022. Pmid:34525713

Iqbal, J, R Little, L Thompson 2021. Using Cover Crops to Reduce Nitrate Leaching in the Waverly Wellhead Protection Area. UNIVERSITY of NEBRASKA–LINCOLN. Available online at: https://cropwatch.unl.edu/2021/using-cover-crops-reduce-nitrate-leaching-waverly-wellhead-protection-area

Isaac, ME, SR Isakson, B Dale et al. 2018. Agroecology in Canada: Towards an integration of agroecological practice, movement, and science. Sustainability, 10, 3299.

Ishtiaque, A 2023. US farmers' adaptations to climate change: A systematic review of adaptation-focused studies in the US agriculture context. Environ. Res. Clim., 2, 022001. https://doi.org/10.1088/2752-5295/accb03

ITF 2019. Efficiency in Railway Operations and Infrastructure Management, ITF Roundtable Reports, 177. OECD Publishing: Paris.

Ivuskin, K, H Bartholomeus, AK Bregt et al. 2019. Global mapping of soil salinity change. Remote Sens. Environ., 231, 111260.

Jablonowski, C, M Herzog, JE Penner et al. 2006. Block-structured adaptive grids on the sphere: Advection experiments. Mon. Weather Rev., 134, 3691–3713.

Jacob, D, J Petersen, B Eggert et al. 2014. EURO-CORDEX: New high-resolution climate change projections for European impact research. Reg. Environ. Change, 14(2), 563–578. https://doi.org/10.1007/s10113-013-0499-2

Jat, ML, D Chakraborty, JK Ladha et al. 2022. Carbon sequestration potential, challenges, and strategies towards climate action in smallholder agricultural systems of South Asia. Crop Environ., 1(1), 86–101.

Jat, ML, Y Singh, G Gill et al. 2014. Laser assisted precision land leveling: Impacts in irrigated intensive production systems of South Asia. Adv. Soil Sci., 2014, 323–352.

Jeffery, S, FGA Verheijen, C Kammann et al. 2016. Biochar effects on methane emissions from soils: A meta-analysis. Soil Biol. Biochem., 101, 251–258.

Jeong, SJ, CH Ho, HJ Gim et al. 2011. Phenology shifts at start vs. end of growing season in temperate vegetation over the northern hemisphere for the period 1982–2008. Glob. Change Biol., 17 (7), 2385–2399. https://doi.org/10.1111/j.1365-2486.2011.02397.x

Jeswani, HK, G Figueroa-Torres, A Azapagic 2021. The extent of food waste generation in the UK and its environmental impacts. Sustain. Prod. Consum., 26, 532–547. Jeswani https://doi.org/10.1016/j.spc.2020.12.021.

Jiang, B, X Deng, H Chen et al. 2023. Research trends and hotspots in climate adaptation of the agricultural system: A bibliometric analysis. Front. Sustain. Food Syst., 7, 1158904. https://doi.org/10.3389/fsufs.2023.1158904

Jiang, J, T Zhou, X Chen et al. 2020. Future changes in precipitation over Central Asia based on CMIP6 projections. Environ. Res. Lett., 15, 1–4.

Jin, Z, SV Archontoulis, DB Lobell 2019. How much will precision nitrogen management pay off? An evaluation based on simulating thousands of corn fields over the US Corn-Belt. Field Crops Res., 240, 12–22. https://doi.org/10.1016/j.fcr.2019.04.013

Jindal, R, B Swallow, J Kerr 2008. Forestry based carbon sequestration projects in Africa: Potential benefits and challenges. Nat. Resour. Forum, 32, 116–130.

Jordaan, SM, E Romo-Rabago, R McLeary et al. 2017. The role of energy technology innovation in reducing greenhouse gas emissions: A case study of Canada. Renew. Sustain. Energy Rev., 78, 1397–1409.

Jose, S, S Bardhan 2012. Agroforestry for biomass production and carbon sequestration: An overview. Agrofor. Syst., 86(2), 105–111.

Joseph, K, C Viswanathan, J Trakler et al. 2003. Regional networking for sustainable landfill management in Asia. In: Proceedings of the Sustainable Landfill Management Workshop, 3–5 December 2003, Anna University, Chennai, p. 39.

Joshi, P 2005. Crop diversification in India: nature, pattern and drivers. In: New Delhi India: Asian Development Bank.

Joshi, VR, MJ Kazula, JA Coulter et al. 2021. In-season weather data provide reliable yield estimates of maize and soybean in the US central Corn Belt. Int. J. Biometeorol., 65, 489–502. https://doi.org/10.1007/s00484-020-02039-z

Jung, S, T Lee, K Gasic et al. 2021. The breeding information management system (BIMS). Database, 2021, baab054. https://doi.org/10.1093/database/baab054

Jury MR 2013. Climate trends in Southern Africa. S. Afr. J. Sci., 109 (1/2). Art. #980, https://doi.org/10.1590/sajs.2013/980

Justes, E 2017. Cover Crops for Sustainable Farming. Dordrecht: Springer. https://doi.org/10.1007/978-94-024-0986-4

Kabir, Z, RT Koide 2002. Effect of autumn and winter mycorrhizal cover crops on soil properties, nutrient uptake and yield of sweet corn in Pennsylvania, USA. Plant Soil, 238, 205–215.

Kakabouki, I, A Tataridas, A Mavroeidis et al. 2021. Introduction of alternative crops in the Mediterranean to satisfy EU Green Deal goals. A review. Agron. Sustain. Dev., 41, 71. https://doi.org/10.1007/s13593-021-00725-9

Kang, S, W Shi, H Cao et al. 2002. Alternate watering in soil vertical profile improved water use efficiency of maize (Zea mays). Field Crops Res., 77, 31–41. https://doi.org/10.1016/S0378-4290(02)00047-3

Kareinen, E, V Uusitalo, A Kuokkanen et al. 2022. Effects of COVID-19 on mobility GHG emissions: Case of the city of Lahti, Finland. Case Stud. Transp. Policy, 10(1), 598–605.

Karl, T, N Nicholls, A Ghazi 1999. Weather and Climate Extremes: Changes, Variations and a Perspective From the Insurance Industry. Dordrecht: Kluwer Academic Publishers.

Karl, TR, RW Knight 1998. Secular trends of precipitation amount, frequency, and intensity in the United States. Bull. Am. Meteorol. Soc., 79(2), 231–241.

Karmakar, R, I Das, D Dutta et al. 2016. Potential effects of climate change on soil properties: A review. Sci. Int., 4, 51–73.

Kaur, G, L Luo, G Chen, et al. 2019. Integrated food waste and sewage treatment—a better approach than conventional food waste-sludge co-digestion for higher energy recovery via anaerobic digestion. Biores. Technol., 289, 121698. https://doi.org/10.1016/j.biortech.2019.121698

Kay, S, C Rega, G Moreno et al. 2019. Agroforestry creates carbon sinks whilst enhancing the environment in agricultural landscapes in Europe. Land Use Policy, 83, 581–593.

Kaye, JP, M Quemada 2017. Using cover crops to mitigate and adapt to climate change. A review. Agron. Sustain. Dev., 37, 4. https://doi.org/10.1007/s13593-016-0410-x

Keating, BA, PS Carberry, GL Hammer et al. 2003. An overview of APSIM, a model designed for farming systems simulation. Eur. J. Agron., 18, 267–288. https://doi.org/10.1016/S1161-0301(02)00108-9

Kebede, AS, RJ Nicholls, A Allan et al. 2018. Applying the Global RCP-SSP-SPA scenario framework at sub-national scale: A multi-scale and participatory scenario approach. Sci. Total Environ., 635, 659–672. https://doi.org/10.1016/j.scitotenv.2018.03.368

Keel, SG, T Anken, L Büchi et al. 2019. Loss of soil organic carbon in Swiss long-term agricultural experiments over a wide range of management practices. Agric. Ecosyst. Environ., 286, 106654. https://doi.org/10.1016/j.agee.2019.106654

Kendon, EJ, RA Stratton, S Tucker et al. 2019. Enhanced future changes in wet and dry extremes over Africa at convection-permitting scale. Nat. Commun., 10, 1794.

Keren, R 2005. Salt-affected soils, reclamation. In Encyclopedia of Soils in the Environment, ed. Hillel, D. (Oxford: Elsevier), pp. 454–461. https://doi.org/10.1016/b0-12-348530-4/00503-8

Kerr, A, J Dialesandro, K Steenwerth et al. 2017. Vulnerability of California specialty crops to projected mid-century temperature changes. Clim. Change, 1–18.

Kerr, D, H Mellon 2012. Energy, population and the environment: Exploring Canada's record on CO_2 emissions and energy use relative to other OECD countries. Popul. Environ., 34, 257–278. https://doi.org/10.1007/s11111-011-0160-2

Kerr, A., J. Dialesandro, K. Steenwerth et al. 2018. Vulnerability of California specialty crops to projected mid-century temperature changes. Climatic Change 148, 419–436 (2018). https://doi.org/10.1007/s10584-017-2011-3

Kettlewell, P, R Byrne, S Jeffery 2023. Wheat area expansion into northern higher latitudes and global food security. Agric. Ecosyst. Environ., 351 (1), 108499. https://doi.org/10.1016/j.agee.2023.108499

Khadka, D, A Aryal, KP Bhatta et al. 2021. Agroforestry systems and their contribution to supplying forest products to communities in the Chure range, Central Nepal. Forests, 12, 358.

Khakbazan, M, K Liu, M Bandara et al. 2022. Pulse-included diverse crop rotations improved the systems economic profitability: Evidenced in two 4-year cycles of rotation experiments. Agron. Sustain. Dev., 42, 103. https://doi.org/10.1007/s13593-022-00831-2

Khalil, MI, B Osborne 2022. Developing climate-resilient agri-environmental production systems. Nutr. Cycl. Agroecosyst., 123, 1–4. https://doi.org/10.1007/s10705-022-10215-3

Khan, Y, F Liu 2023. Consumption of energy from conventional sources a challenge to the green environment: Evaluating the role of energy imports, and energy intensity in Australia. Environ. Sci. Pollut. Res., 30, 22712–22727. https://doi.org/10.1007/s11356-022-23750-x

Khapayi, M, PR Celliers 2016. Factors limiting and preventing emerging farmers to progress to commercial agricultural farming in the King William's town area of the Eastern Cape Province, South Africa. SAJAE, 44(1), 25–41. https://www.ajol.info/index.php/sajae/article/view/138546

Kichey, T, B Hirel, E Heumez et al. 2007. Winter wheat (*Triticum aestivum* L.), post-anthesis nitrogen uptake and remobilization to the grain correlates with agronomic traits and nitrogen physiological markers. Field Crop. Res., 102, 22–32.

Kijne, JW, R Barker, D Molden 2003. Water Productivity in Agricultural: Limits and Opportunities for Improvement. Wallingford: CAB International.

Kim, DG, E Grieco, A Bombelli et al. 2021a. Challenges and opportunities for enhancing food security and greenhouse gas mitigation in smallholder farming in sub-Saharan Africa. A review. Food Secur. https://doi.org/10.1007/s12571-021-01149-9

Kim, DG, E Grieco, A Bombelli et al. 2021b. Challenges and opportunities for enhancing food security and greenhouse gas mitigation in smallholder farming in sub-Saharan Africa. A review. Food Secur. https://doi.org/10.1007/s12571-021-01149-9

Kim, DG, MUF Kirschbaumb, TL Beedy 2016. Carbon sequestration and greenhouse gas emissions in agroforestry: Summary of available data and suggestion for future studies. Agric. Ecosyst. Environ., 226, 65–78. https://doi.org/10.1016/j.agee.2016.04.011

Kim IW, J Oh, S Woo et al. 2019. Evaluation of precipitation extremes over the Asian domain: Observation and modelling studies. Clim. Dyn., 52, 1317–1342. https://doi.org/10.1007/s00382-018-4193-4

Kimaro, EG, SM Mor, JALML Toribio 2018. Climate change perception and impacts on cattle production in pastoral communities of northern Tanzania. Pastoralism, 8, 19. https://doi.org/10.1186/s13570-018-0125-5

Kimathi, SM, OI Ayuya, B Mutai 2021. Adoption of climate-resilient potato varieties under partial population exposure and its determinants: Case of smallholder farmers in Meru County, Kenya. Cogent Food Agric., 7, 1860185.

King, M, D Altdorff, P Li et al. 2018. Northward shift of the agricultural climate zone under 21st-century global climate change. Sci. Rep., 8(1), 1–10. https://doi.org/10.1038/s41598-018-26321-8

Kiss, B, Z Szalay 2022. Sensitivity of buildings' carbon footprint to electricity decarbonization: A life cycle–based multi-objective optimization approach. Int. J. Life Cycle Assess. https://doi.org/10.1007/s11367-022-02043-y

Kiyani, P, J Andoh, Y Lee et al. 2017. Benefits and challenges of agroforestry adoption: A case of Musebeya sector, Nyamagabe District in southern province of Rwanda. For. Sci. Technol., 13(4), 174–180.

Kleanthis, N, V Stavrakas, A Ceglarz et al. 2022. Eliciting knowledge from stakeholders to identify critical issues of the transition to climate neutrality in Greece, the Nordic Region, and the European Union. Energy Res. Soc. Sci., 93, 102836.

Klein, I, U Gessner, C Kuenzer 2012. Regional land cover mapping and change detection in Central Asia using MODIS time-series. Appl. Geogr., 35(1–2), 219–234.

Knox, J, A Daccache, T Hess et al. 2016. Meta-analysis of climate impacts and uncertainty on crop yields in Europe. Environ. Res. Lett., 11, 113004.

Knox, J, T Hess, A Daccache et al. 2012. Climate change impacts on crop productivity in Africa and South Asia. Environ. Res. Lett., 7, 034032. https://doi.org/10.1088/1748-9326/7/3/034032

Kogo, BK, L Kumar, R Koech et al. 2019. Modelling impacts of climate change on maize (*Zea mays* L.) Growth and productivity: A review of models, outputs and limitations. J. Geosci. Environ. Prot., 7, 76.

Konapala, G, AK Mishra, Y Wada et al. 2020. Climate change will affect global water availability through compounding changes in seasonal precipitation and evaporation. Nat. Commun., 11, 1–10.

Kothari, K, S Ale, GW Marek et al. 2022. Simulating the climate change impacts and evaluating potential adaptation strategies for irrigated corn production in Northern High Plains of Texas. Clim. Risk Manag., 37, 100446.

Koudahe, K, SC Allen, K Djaman 2022. Critical review of the impact of cover crops on soil properties. Int. Soil Water Conserv., 10, 343–354.

Kovats, S, S Lloyd, A Hunt et al. 2011. Health: the Impacts and Economic Costs of Climate Change on Health in Europe. Oxford: Stockholm Environment Institute.

Kreidenweis, U, J Breier, C Herrmann et al. 2021. Greenhouse gas emissions from broiler manure treatment options are lowest in well-managed biogas production. J. Clean. Prod., 280, 124969.

Kristanto, GA, W Koven 2019. Estimating greenhouse gas emissions from municipal solid waste management in Depok, Indonesia. City Environ. Interact., 4, 100027.

Kritee, K, D Nair, D Zavala-Araiza et al. 2018. High nitrous oxide fluxes from rice indicate the need to manage water for both long- and short-term climate impacts. Proc. Natl. Acad. Sci., 115, 9720–9725. https://doi.org/10.1073/pnas.1809276115

Kruse, J 2011. Estimating Demand for Agricultural Commodities to 2050; Report No. 3-16-10; Global Harvest Initiative: Washington, DC, USA.

Kulshreshtha, S 2019. Resiliency of Prairie Agriculture to Climate Change. IntechOpen. https://doi.org/10.5772/intechopen.87098

Kulshreshtha, SN 2011. Climate change, prairie agriculture, and prairie economy: The new normal. Can. J. Agric. Econ., 59, 19–44.

Kumar, A, A Raman, S Yadav et al. 2021. Genetic gain for rice yield in rainfed environments in India. Field Crops Res., 260, 107977. https://doi.org/10.1016/j.fcr.2020.107977

Kumar, M, Z Shao, C Braun et al. 2022. Decarbonizing India's road transport: A meta-analysis of road transport emissions models. ICCT WHITE PAPER.

Kuyah, S, CW Whitney, M Jonsson et al. 2019. Agroforestry delivers a win-win solution for ecosystem services in sub-Saharan Africa. A meta-analysis. Agron. Sustain. Dev., 39, 47. https://doi.org/10.1007/s13593-019-0589-8

Kwasny T, K Dobernig, P Riefler 2022. Towards reduced meat consumption: A systematic literature review of intervention effectiveness, 2001–2019. Appetite, 168, 105739. https://doi.org/10.1016/j.appet.2021.105739

Kwesiga, J, K Grotelüschen, D Neuhoff et al. 2019. Site and management effects on grain yield and yield variability of rainfed lowland rice in the Kilombero floodplain of Tanzania. Agronomy, 9(10), 632. https://doi.org/10.3390/agronomy9100632.

Laatsch, J, Z Ma 2015. Strategies for incorporating climate change into public forest management. J. For., 113, 335–342.

Lacombe, G, P Chinnasamy, A Nicol 2019. Review of climate change science, knowledge and impacts on water resources in South Asia. Background Paper 1. Colombo, Sri Lanka: International Water Management Institute (IWMI), 73. (Climate Risks and Solutions: Adaptation Frameworks for Water Resources Planning, Development and Management in South Asia). https://doi.org/10.5337/2019.202, http://www.iwmi.cgiar.org/Publications/Other/PDF/sawi-paper-1.pdf

Laforge, J, V Corkal, A Cosbey 2021. Farming the Future: Agriculture and Climate Change on the Canadian Prairies. The International Institute for Sustainable Development. https://www.iisd.org/system/files/2021-11/farming-future-agriculture-climate-change-canadian-prairies.pdf

Laforge JML, B Dale, CZ Levkoe et al. 2021. The future of agroecology in Canada: Embracing the politics of food sovereignty. J. Rural Stud., 81, 194–202. https://doi.org/10.1016/j.jrurstud.2020.10.025.

Lal, R 2004. Soil carbon sequestration to mitigate climate change. Geoderma, 123, 1–22.

Lal R 2017. Soil erosion by wind and water: problems and prospects Soil Erosion Research Methods, Routledge, pp. 1–10, https://doi.org/10.1201/9780203739358-1

Lal, R, MVK Sivakumar, SMA Faiz et al. 2011. Climate Change and Food Security in South Asia. Springer Science and Business Media B.V., pp. 137–152.

Lam, VWY, WWL Cheung, G Reygondeau et al. 2016. Projected change in global fisheries revenues under climate change. Sci. Rep., 6, 32607. https://doi.org/10.1038/srep32607

Lamb, WF, T Wiedmann, J Pongratz et al. 2021. A review of trends and drivers of greenhouse gas emissions by sector from 1990 to 2018. Environ. Res. Lett., 16(7), 073005. https://doi.org/10.1088/1748-9326/ABEE4E

Lansigan, FP, WL De los Santos, JO Coladilla 2000. Agronomic impacts of climate variability on rice production in the Philippines. Agric. Ecosyst. Environ., 82(1–3), 129–137.

Lassaletta, L, G Billen, B Grizzetti et al. 2014. 50 year trends in nitrogen use efficiency of world cropping systems: The relationship between yield and nitrogen input to cropland. Environ. Res. Lett., 9, 105011. https://doi.org/10.1088/1748-9326/9/10/105011.

Latif, M 2011. Uncertainty in climate change projections. J. Geochem. Explor., 110(1), 1–7. https://doi.org/10.1016/j.gexplo.2010.09.011

Le Cozannet, G, JC Manceau, J Rohmer 2017. Bounding probabilistic sea-level projections within the framework of the possibility theory. Environ. Res. Lett., 12, 014012. https://doi.org/10.1088/1748-9326/aa5528

Leal Filho, W, E Totin, JA Franke et al. 2022. Understanding responses to climate-related water scarcity in Africa. Sci. Total Environ., 806, 150420, https://doi.org/10.1016/j.scitotenv.2021.150420

Lebel, S, L Fleskens, PM Forster et al. 2015. Evaluation of in situ rainwater harvesting as an adaptation strategy to climate change for maize production in rainfed Africa. Water Resour. Manag., 29, 4803–4816.

Leclère, D, M Obersteiner, M Barrett et al. 2020. Bending the curve of terrestrial biodiversity needs an integrated strategy. Nature, 585, 551–556.

Lee, DK, DH Cha 2020. Regional climate modeling for Asia. Geosci. Lett., 7, 13. https://doi.org/10.1186/s40562-020-00162-8

Lee, H, D Sumner 2015. Economics of downscaled climate-induced changes in cropland, with projections to 2050: Evidence from Yolo County California. Clim. Change, 132, 723–737. https://doi.org/10.1007/s10584-015-1436-9

Lee, J, S De Gryze, J Six 2011. Effect of climate change on field crop production in California's Central Valley. Clim. Change, 109, 335–353.

Lee, JW, EM Hong, JU Kim et al. 2022. Evaluation of agricultural drought in South Korea using socio-economic drought information. Int. J. Disaster Risk Reduct., 74, 102936.

Lehmkuhl, F, H Schüttrumpf, J Schwarzbauer et al. 2022. Assessment of the 2021 summer flood in Central Europe. Environ. Sci. Eur., 34, 107. https://doi.org/10.1186/s12302-022-00685-1

Lehner, F, C Deser, N Maher et al. 2020. Partitioning climate projection uncertainty with multiple large ensembles and CMIP5/6. Earth Syst. Dynam., 11, 491–508, https://doi.org/10.5194/esd-11-491-2020

Lehner, F, S Coats, TF Stocker et al. 2017. Projected drought risk in 1.5 °C and 2 °C warmer climates. Geophys. Res. Lett., 44, 7419–7428.

Leip, A, G Billen, J Garnier et al. 2015. Impacts of European livestock production: Nitrogen, sulphur, phosphorus and greenhouse gas emissions, land-use, water eutrophication and biodiversity. Environ. Res. Lett., 10, 115004.

Lemieux, CJ, TJ Beechey, DJ Scott et al. 2010. Protected Areas and Climate Change in Canada: Challenges and Opportunities for Adaptation. Technical Report No. 19. Canadian Council on Ecological Areas, Ottawa, Ontario.

Lemieux, CJ, TJ Beechey, PA Gray 2011. Prospects for Canada's protected areas in an era of rapid climate change. Land Use Policy, 28(4), 928–941. https://doi.org/10.1016/j.landusepol.2011.03.008

Leng, G, J Hall 2019. Crop yield sensitivity of global major agricultural countries to droughts and the projected changes in the future. Sci. Total Environ., 654, 811–821. https://doi.org/10.1016/j.scitotenv.2018.10.434

Lesschen, JP, M van den Berg, HJ Westhoek et al. 2011. Oenema Greenhouse gas emission profiles of European livestock sectors. Anim. Feed Sci. Technol., 166–167, 16–28.

Letseku, V, B Grové 2022. Crop water productivity, applied water productivity and economic decision making. Water, 14, 1598. https://doi.org/10.3390/w14101598

Levin, K. 2018. 8 Things You Need to Know About the IPCC 1.5°C Report October World Resources Institute (2018). Retrieved from https://www.wri.org/blog/2018/10/8-things-you-need-know-about-ipcc-15-c-report

Leweri, CM, MJ Msuha, AC Treydte 2021. Rainfall variability and socio-economic constraints on livestock production in the Ngorongoro Conservation Area, Tanzania. SN Appl. Sci., 3, 123. https://doi.org/10.1007/s42452-020-04111-0

Li, G, X Zhang, AJ Cannon et al. 2018. Indices of Canada's future climate for general and agricultural adaptation applications. Clim. Change, 148, 249–263. https://doi.org/10.1007/s10584-018-2199-x

Li, S, L Chen, X Han et al. 2022. Rice cultivar renewal reduces methane emissions by improving root traits and optimizing photosynthetic carbon allocation. Agriculture, 12, 2134. https://doi.org/10.3390/agriculture12122134

Li W, H Li, H Zhang et al. 2016. The analysis of CO_2 emissions and reduction potential in China's transport sector. Math. Probl. Eng., 2016, 1043717. https://doi.org/10.1155/2016/1043717

Li X, D Long, BR Scanlon et al. 2022. Climate change threatens terrestrial water storage over the Tibetan Plateau. Nat. Clim. Change, 12, 801–807. https://doi.org/10.1038/s41558-022-01443-0

Li Z, C Zhan, S Hu, et al. 2022. Evaluation of global gridded crop models (GGCMs) for the simulation of major grain crop yields in China. Hydrol. Res., 53 (3), 353–369. https://doi.org/10.2166/nh.2022.087

Li, Z, H Fang 2016. Impacts of climate change on water erosion: A review. Earth-Sci. Rev., 163, 94–117.

Liang, D, X Lu, M Zhuang et al. 2021. China's greenhouse gas emissions for cropping systems from 1978–2016. Sci. Data, 8, 171. https://doi.org/10.1038/s41597-021-00960-5

Liang, X, Z Coauthors 2019. CWRF performance at downscaling China climate characteristics. Clim. Dyn., 52, 2159–2184.

Lin, BB 2011. Resilience in agriculture through crop diversification: Adaptive management for environmental change. Bioscience, 61, 183–193.

Lin, T, DC Catacutan, M van Noordwijk et al. 2021. State and Outlook of Agroforestry in ASEAN: Status, trends and Outlook 2030 and Beyond. Food and Agriculture Organization of the United Nations. 2021. Regional Office for Asia and the Pacific.

Liu, H, HH Li, HF Ning et al. 2019. Optimizing irrigation frequency and amount to balance yield, fruit quality and water use efficiency of greenhouse tomato. Agric. Water Manag., 226, 105787. https://doi.org/10.1016/j.agwat.2019.105787

Liu, J, D Chen, G Mao et al. 2021. Past and future changes in climate and water resources in the Lancang–Mekong River Basin: Current understanding and future research directions. Engineering, 13, 144–152. https://doi.org/10.1016/j.eng.2021.06.026

Liu, J, M Wang, L Yang et al. 2020. Agricultural productivity growth and its determinants in South and Southeast Asian countries. Sustainability, 12(12), 4981. https://doi.org/10.3390/su12124981

Liu, L, B Basso 2020. Impacts of climate variability and adaptation strategies on crop yields and soil organic carbon in the US Midwest. PLOS One, 15(1), e0225433. https://doi.org/10.1371/journal.pone.0225433

Liu, LL, EL Wang, Y Zhu et al. 2012. Contrasting effects of warming and autonomous breeding on single-rice productivity in China. Agric. Ecosyst. Environ., 149, 20–29. https://doi.org/10.1016/j.agee.2011.12.008

Liu Q, J Niu, B Sivakumar et al. 2021. Accessing future crop yield and crop water productivity over the Heihe River basin in northwest China under a changing climate. Geosci. Lett., 8, 2. https://doi.org/10.1186/s40562-020-00172-6

Liu, Y, Y Liu, W Wang et al. 2022. Soil moisture droughts in East Africa: Spatiotemporal patterns and climate drivers. J. Hydrol. Reg. Stud., 40, 101013. https://doi.org/10.1016/j.ejrh.2022.101013

Liu, Z 2015. Global Energy Interconnection, 1st edition, Academic Press.

Liu, Z, P Ciais, Z Deng et al. 2020. Near-real-time monitoring of global CO_2 emissions reveals the effects of the COVID-19 pandemic. Nat. Commun., 11, 5172.

Lobell, DB, MB Burke 2010. On the use of statistical models to predict crop yield responses to climate change. Agric. For. Meteorol., 150, 1443–1452.

Lobell, DB, MJ Roberts, W Schlenker et al. 2014. Greater sensitivity to drought accompanies maize yield increase in the U.S. Midwest. Science, 344(6183), 516–519. https://doi.org/10.1126/science.1251423

Lobell, DB, W Schlenker, J Costa-Roberts 2011. Climate trends and global crop production since 1980. Science, 333, 616–620.

Long, SP, EA Ainsworth, ADB Leakey et al. 2006. Food for thought: Lower-than-expected crop yield stimulation with rising CO_2 concentrations. Science, 312, 1918–1921.

López-Ballesteros, A, J Beck, A Bombelli et al. 2018: Towards a feasible and representative pan-African research infrastructure network for GHG observations. Environ. Res. Lett., 13, 085003. https://doi.org/10.1088/1748-9326/aad66c

Low-Carbon Asia Research Project 2013. Realizing low carbon Asia, contribution of ten actions. Full report is available at: http://2050.nies.go.jp/file/ten_actions_2013.pdf

Lu, WC 2017. Greenhouse gas emissions, energy consumption and economic growth: A panel cointegration analysis for 16 Asian countries. Int. J. Environ. Res. Public Health, 14(11), 1436. https://doi.org/10.3390/ijerph14111436

Luo, L, EF Wood 2007. Monitoring and predicting the 2007 U.S. drought. Geophys. Res. Lett., 34, L22702. https://doi.org/10.1029/2007GL031673

Łupikasza, EB 2017. Seasonal patterns and consistency of extreme precipitation trends in Europe, December 1950 to February 2008. Clim. Res., 72, 217–237. https://doi.org/10.3354/cr01467

Lychuk, TE, AP Moulin, RL Lemke et al. 2019. Climate change, agricultural inputs, cropping diversity, and environment affect soil carbon and respiration: A case study in Saskatchewan, Canada. Geoderma, 337, 664–678.

Lyon, B, DG Dewitt 2012. A recent and abrupt decline in the East African long rains. Geophys. Res. Lett. https://doi.org/10.1029/2011GL050337

Ma, Z, EW Bork, CN Carlyle et al. 2022. Carbon stocks differ among land-uses in agroforestry systems in western Canada. Agric. For. Meteorol., 313, 108756.

Ma, Z, HYH Chen, EW Bork et al. 2020. Carbon accumulation in agroforestry systems is affected by tree species diversity, age and regional climate: A global meta-analysis. Glob. Ecol. Biogeogr., 29, 1817–1828, https://doi.org/10.1111/geb.13145

MacCarthy, DS, PS Traore, BS Freduah et al. 2022. Productivity of soybean under projected climate change in a semi-arid region of West Africa: Sensitivity of current production system. Agronomy, 12, 2614. https://doi.org/10.3390/agronomy12112614

Magiri, R, K Muzandu, G Gitau, et al. 2020. Impact of Climate Change on Animal Health, Emerging and Re-emerging Diseases in Africa. African Handbook of Climate Change Adaptation, pp. 1–17.

Mahlalela, PT, RC Blamey, NCG Hart et al. 2020. Drought in the Eastern Cape region of South Africa and trends in rainfall characteristics. Clim. Dyn., 55, 2743–2759. https://doi.org/10.1007/s00382-020-05413-0

Makate, C, R Wang, M Makate et al. 2016. Crop diversification and livelihoods of smallholder farmers in Zimbabwe: Adaptive management for environmental change. SpringerPlus, 5, 1135. https://doi.org/10.1186/s40064-016-2802-4

Malek, K, J Adam, C Stockle et al. 2018. When should irrigators invest in more water-efficient technologies as an adaptation to climate change? Water Resour. Res., 54, 8999–9032.

Malhi, GS, M Kaur, P Kaushik 2021. Impact of climate change on agriculture and its mitigation strategies: A review. Sustainability, 13, 1318. https://doi.org/10.3390/su13031318

Mălinaş, A, R Vidican, I Rotar et al. 2022. Current status and future prospective for nitrogen use efficiency in wheat (*Triticum aestivum* L.). Plants (Basel), 11(2), 217. https://doi.org/10.3390/plants11020217

Manaye, A, M Negash, M Alebachew 2019. Effect of degraded land rehabilitation on carbon stocks and biodiversity in semi-arid region of Northern Ethiopia. Forest Sci. Technol., 15, 70–79.

Manning, B, F Pollinger, A Gafurov et al. 2018. Impacts of climate change in central Asia. Encycl. Anthr., 195–203. https://doi.org/10.1016/B978-0-12-809665-9.09751-2

Marcos, M, MN Tsimplis, AGP Shaw 2009. Sea level extremes in southern Europe. J. Geophys. Res. Oceans, 114, C01007.

Mardian, J 2022. The role of spatial scale in drought monitoring and early warning systems: A review. Environ. Rev., 30(3), 438–459. https://doi.org/10.1139/er-2021-0102

Martey E, PM Etwire, JKM Kuwornu 2020. Economic impacts of smallholder farmers' adoption of drought-tolerant maize varieties. Land Use Policy, 94, 104524. https://doi.org/10.1016/j.landusepol.2020.104524

Martínez, MM, T Nakaegawa, R Pinzón et al. 2020. Using a statistical crop model to predict maize yield by the end-of-century for the Azuero Region in Panama. Atmosphere, 11, 1097. https://doi.org/10.3390/atmos11101097

Marx, A, A Bastrup-Birk, G Louwagie et al. 2017. Terrestrial ecosystems, soil and forests. In: Climate Change, Impacts and Vulnerability in Europe 2016 – An Indicator-based Report. https://doi.org/10.1007/978-3-540-88246-6

Marzeion, B, AH Jarosch, M Hofer 2012. Past and future sea-level change from the surface mass balance of glaciers. Cryosphere, 6, 1295–1322. https://doi.org/10.5194/tc-6-1295-2012

Masui, T, S Ashina, S Fujimori et al. 2016. GHG Reduction Potential in Asia. In: Nishioka S (eds.), Enabling Asia to Stabilise the Climate. Springer, Singapore. https://doi.org/10.1007/978-981-287-826-7_1

Maurya, K, S Mondal, V Kumar et al. 2021. Roadmap to sustainable carbon-neutral energy and environment: Can we cross the barrier of biomass productivity? Environ. Sci. Pollut. Res., 28, 49327–49342. https://doi.org/10.1007/s11356-021-15540-8

May, WE, RM Mohr, GP Lafond et al. 2004. Early seeding dates improve oat yield and quality in the eastern prairies. Can. J. Plan. Sci., 84, 431–442.

Mayer, S, M Wiesmeier, E Sakamoto et al. 2022. Soil organic carbon sequestration in temperate agroforestry systems a meta-analysis. Agric. Ecosyst. Environ., 323, 107689, https://doi.org/10.1016/j.agee.2021.107689

Mbow, C, M van Noordwijk, E Luedeling et al. 2014. Agroforestry solutions to address food security and climate change challenges in Africa. Curr. Opin. Environ. Sustain., 6, 61–67, https://doi.org/10.1016/j.cosust.2013.10.014

McDonald, RI, EH Girvetz 2013. Two challenges for U.S. irrigation due to climate change: Increasing irrigated area in wet states and Increasing Irrigation rates in dry states. PLOS One, 8(6), e65589. https://doi.org/10.1371/journal.pone.0065589

McKinsey Global Institute 2020. Climate Risk and Response in Asia. McKInsey Global Institute.

McLachlan, A 2019. Africa, Coastal Ecology. In: Finkl, C.W., Makowski, C. (eds.), Encyclopedia of Coastal Science. Encyclopedia of Earth Sciences Series. Springer, Cham. https://doi.org/10.1007/978-3-319-93806-6_2

Medellín-Azuara, J, DA Sumner, QY Pan et al. 2018. Economic and Environmental Implications of California Crop and Livestock, Adaptation to Climate Change. California Natural Resources Agency. Publication number: CCCA4-CNRA-2018-018.

Meessen, J, J Pestiaux, M Cornet et al. 2020. Increasing the EU's 2030 Emissions Reduction Target: How to Cut EU GHG Emissions by 55 Percent or 65 Percent by 2030. Climact, 51.

Meinshausen, M, J Lewis, C McGlade et al. 2022. Realization of Paris agreement pledges may limit warming just below 2°C. Nature, 604(7905), 304–309.

Mekonen, AA, AB Berlie 2020. Spatiotemporal variability and trends of rainfall and temperature in the Northeastern highlands of Ethiopia. Model. Earth Syst. Environ., 6, 285–300. https://doi.org/10.1007/s40808-019-00678-9

Mekuria, W, A Noble, M McCartney et al. 2015. Soil management for raising crop water productivity in rainfed production systems in Lao PDR. Arch. Agron. Soil Sci., 62(1), 53–68. https://doi.org/10.1080/03650340.2015.1037297

Melillo, JM, SD Frey, KM DeAngelis et al. 2017. Long-term pattern and magnitude of soil carbon feedback to the climate system in a warming world. Science, 358, 101–105. https://doi.org/10.1126/science.aan2874

Melillo, JM, TC Richmond, GW Yohe 2014. Highlights of climate change impacts in the United States: The third national climate assessment. US Congress.

Meltzer, MI 1995. Livestock in Africa: The economics of ownership and production, and the potential for improvement. Agric. Hum. Values, 12, 4–18. https://doi.org/10.1007/BF02217292

Mendelsohn, R 2014. The impact of climate change on agriculture in Asia. J. Integr. Agric., 13(4), 660–665.

Merbold, L, RJ Scholes, M Acosta et al. 2021. Opportunities for an African greenhouse gas observation system. Reg. Environ. Change, 21, 104. https://doi.org/10.1007/s10113-021-01823-w

Mgalula, ME, OV Wasonga, C Hülsebusch et al. 2021. Greenhouse gas emissions and carbon sink potential in Eastern Africa rangeland ecosystems: A review. Pastoralism, 11, 19. https://doi.org/10.1186/s13570-021-00201-9

Mielcarek-Bocheńska P, W Rzeźnik 2021. Greenhouse gas emissions from agriculture in EU countries—State and perspectives. Atmosphere, 12, 1396. https://doi.org/10.3390/atmos12111396

Migliorini, P, A Wezel 2017. Converging and diverging principles and practices of organic agriculture regulations and agroecology. A review. Agron. Sustain. Dev., 37, 63. https://doi.org/10.1007/s13593-017-0472-4

Minang, PA, LA Duguma, F Bernard et al. 2014. Prospects for agroforestry in REDD+ landscapes in Africa. Curr. Opin. Environ. Sustain., 6, 78–82.

Misganaw Engdaw, M, AP Ballinger, GC Hegerl et al. 2021. Changes in temperature and heat waves over Africa using observational and reanalysis data sets. Int. J. Climatol., https://doi.org/10.1002/joc.7295

Mishra, AK, VP Singh 2010. Changes in extreme precipitation in Texas. J. Geophys. Res. Atmos., 2010, 115.

Mishra, V, JM Wallace, DP Lettenmaier 2012. Relationship between hourly extreme precipitation and local air temperature in the United States. Geophys. Res. Lett., 39, L16403. https://doi.org/10.1029/2012GL052790

Mishra, V, U Bhatia, AD Tiwari 2020. Bias-corrected climate projections for South Asia from coupled model intercomparison project-6. Sci. Data, 7, 338. https://doi.org/10.1038/s41597-020-00681-1

Mohammed, S, K Alsafadi, I Takács et al. 2020. Contemporary changes of greenhouse gases emission from the agricultural sector in the EU-27. Geol. Ecol. Landsc., 4, 282–287.

Mohan, G, R Mishra, AA Reddy et al. 2022. Scaling up micro-irrigation technology to address water challenges in semi-arid south Asia. UNU-IAS Policy Brief series. United Nations University Institute for the Advanced Study of Sustainability. Available online at: https://collections.unu.edu/eserv/UNU:8636/UNU-IAS-PB-No30-2022.pdf

Mohanavelu, A, SR Naganna, N Al-Ansari 2021. Irrigation induced salinity and sodicity hazards on soil and groundwater: An overview of its causes, impacts and mitigation strategies. Agriculture, 11, 983. https://doi.org/10.3390/agriculture11100983

Moir, JL, BJ Malcolm, KC Cameron et al. 2012. The effect of dicyandiamide on pasture nitrate concentration, yield and N off take under high N loading in winter and spring. Grass Forage Sci., 67, 391–402. https://doi.org/10.1111/j.1365-2494.2012.00857.x

Molua, EL, CM Lambi 2007. The economic impact of climate change on agriculture in Cameroon. Policy Research Working Paper; No. 4364. World Bank, Washington, DC. https://openknowledge.worldbank.org/entities/publication/25170c67-55db-5352-be52-805c6b52b3c4 License: CC BY 3.0 IGO.

Momodu, AS, ID Okunade, TD Adepoju 2022. Decarbonising the electric power sectors in sub-Saharan Africa as a climate action: A systematic review. Environ. Chall., 7, 100485.

Monier E, X Gao 2015. Climate change impacts on extreme events in the United States: An uncertainty analysis. Clim. Change, 131, 67–81. https://doi.org/10.1007/s10584-013-1048-1

Moore, FC, DB Lobell 2015. The fingerprint of climate trends on European crop yields. Proc. Natl. Acad. Sci. U. S. A., 112(9), 2670–2675. https://doi.org/10.1073/pnas.1409606112

Morales, MB, M Díaz, D Giralt et al. 2022. Protect European green agricultural policies for future food security. Commun. Earth Environ., 3, 217, https://doi.org/10.1038/s43247-022-00550-2

Morgounov, A, K Sonder, A Abugalieva et al. 2018. Effect of climate change on spring wheat yields in North America and Eurasia in 1981–2015 and implications for breeding. PLOS One, 13(10), e0204932. https://doi.org/10.1371/journal.pone.0204932

Morignat, E, JB Perrin, E Gay et al. 2014. Assessment of the impact of the 2003 and 2006 heat waves on cattle mortality in France. PLOS One, 9, e93176. https://doi.org/10.1371/journal.pone.0093176

Morita 2021. Measure for raising crop water productivity in South Asia and Sub-Saharan Africa, in: M. Dinesh Kumar (ed.), Current Directions in Water Scarcity Research, Elsevier, 3, 157–196, ISSN 2542-7946,ISBN 9780323912778, https://doi.org/10.1016/B978-0-323-91277-8.00011-3

Morrison, MJ, DW Stewart 2002. Heat stress during flowering in summer brassica. Crop Sci., 42, 797–803. https://doi.org/10.2135/cropsci2002.0797

Mosquera-Losada, MR, MGS Santos, B Gonçalves et al. 2023. Policy challenges for agroforestry implementation in Europe. Front. For. Glob. Change, 6, 1127601. https://doi.org/10.3389/ffgc.2023.1127601

Motha, RP, W Baier 2005. Impacts of present and future climate change and climate variability on agriculture in the temperate regions: North America. Clim. Change, 70, 137–164. https://doi.org/10.1007/s10584-005-5940-1

Mottaleb, KA, RM Rejesus, M Murty et al. 2017. Benefits of the development and dissemination of climate-smart rice: Ex ante impact assessment of drought-tolerant rice in South Asia. Mitig. Adapt. Strateg. Glob. Change, 22, 879–901.

Msangi, S, L You 2010. Food Security, Farming and Climate Change to 2050: Scenarios, Results, Policy Options. Research monograph, International Food Policy Research Institute: Washington, DC, USA.

Mu, L, C Wang, B Xue et al. 2019. Assessing the impact of water price reform on farmers' willingness to pay for agricultural water in northwest China. J. Clean. Prod., 234, 1072–1081.

Mu, L, J Janmaat, J Taylor et al. 2023. Attitudes and opportunities: Comparing climate change adaptation intentions and decisions of agricultural producers in Shaanxi, China, and British Columbia, Canada. Mitig. Adapt. Strateg. Glob. Change, 28, 8. https://doi.org/10.1007/s11027-022-10040-7

Muchuru, S, G Nhamo 2019. A review of climate change adaptation measures in the African crop sector. Clim. Dev., 11, 873–885. https://doi.org/10.1080/17565529.2019.1585319

Muhammad, FR, SW Lubis, S Setiawan 2021. Impacts of the Madden–Julian oscillation on precipitation extremes in Indonesia. Int. J. Climatol., 41(3), 1970–1984. https://doi.org/10.1002/joc.6941

Mukhlis, I, MS Rizaludin, I Hidayah 2022. Understanding socio-economic and environmental impacts of agroforestry on rural communities. Forests, 13, 556.

Müller C, J Elliott, J Chryssanthacopoulos et al. 2017. Global gridded crop model evaluation: Benchmarking, skills, deficiencies and implications. Geosci. Model Dev., 10, 1403–1422. https://doi.org/10.5194/gmd-10-1403-2017

Müller, C, RD Robertson 2014. Projecting future crop productivity for global economic modeling. Agric. Econ., 45, 37–50.

Mulungu, K, JN Ng'ombe 2020. Climate change impacts on sustainable maize production in Sub-Saharan Africa: A review. Maize Prod. Use. https://doi.org/10.5772/intechopen.90033

Muthoni, J, DN Nyamongo, M Mbiyu 2017. Climatic change, its likely impact on potato (*Solanum tuberosum* L.) production in Kenya and plausible coping measures. Int. J. Hortic., 7, 115–123.

Nalumu, DJ, H Mensah, O Amponsah et al. 2021. Stakeholder collaboration and irrigation practices in Ghana: Issues, challenges, and the way forward. SN Appl. Sci., 3, 576. https://doi.org/10.1007/s42452-021-04407-9

Nandan, R, V Singh, SS Singh et al. 2019. Impact of conservation tillage in rice–based cropping systems on soil aggregation, carbon pools and nutrients. Geoderma, 340, 104–114.

Nangombe, SS, T Zhou, W Zhang et al. 2019. High-temperature extreme events over Africa under 1.5°C and 2°C of global warming. J. Geophys. Res.: Atmos., 124(8), 4413–4428. https://doi.org/10.1029/2018JD029747

Narbayep, M, V Pavlova 2022. The Aral Sea, Central Asian Countries and Climate Change in the 21st Century. United Nations ESCAP, IDD, April 2022. Bangkok. Available at: http://www.unescap.org/kp

NASA 2020. Carbon Dioxide. https://climate.nasa.gov/vital-signs/carbon-dioxide/. Published November 2020.

Nascimento, CM, W de Sousa Mendes, NEQ Silvero et al. 2021. Soil degradation index developed by multitemporal remote sensing images, climate variables, terrain and soil attributes. J. Environ. Manag., 277, 111316. https://doi.org/10.1016/j.jenvman.2020.111316

Nascimento, L, T Kuramochi, N Höhne 2022. The G20 emission projections to 2030 improved since the Paris Agreement, but only slightly. Mitig. Adapt. Strateg. Glob. Change, 27, 39.

National Bureau of Statistics 2018. China Energy Statistics Yearbook 2018. China Statistics Press.

Natural Resources Canada 2017. About Renewable Energy. Natural Resources Canada Available at: https://natural-resources.canada.ca/our-natural-resources/energy-sources-distribution/renewable-energy/about-renewable-energy/7295

Naumann, G, C Cammalleri, L Mentaschi et al. 2021. Increased economic drought impacts in Europe with anthropogenic warming Nat. Clim. Change, 11, 485–491. https://doi.org/10.1038/s41558-021-01044-3

Naumann, G, L Alfieri, K Wyser et al. 2018. Global changes in drought conditions under different levels of warming. Geophys. Res. Lett., 45, 3285–3296. https://doi.org/10.1002/2017GL076521

Naveendrakumar, G, M Vithanage, HH Kwon et al. 2019. South Asian Perspective on temperature and rainfall extremes: A review. Atmos. Res., 225, 110–120.

Navius Research 2021. Achieving net zero emissions by 2050 in Canada. An evaluation of pathways to net zero prepared for the Canadian Institute for Climate Choices. Navius Research, https://climatechoices.ca/wp-content/uploads/2021/02/Deep-Decarbonization-Report-2021-01-21-FINAL_EN.pdf

Tahroudi, MN, HK Nejad 2017. Evaluation the trend of South-West of Asia precipitations. Water Soil, 31(5), 1511–1525. https://doi.org/10.22067/jsw.v31i5.62604

Nearing, MA, FF Pruski, MR O'Neal 2004. Expected climate change impacts on soil erosion rates: A review. J. Soil Water Conserv., 59(1), 43–50.

Nelson, GC, H Valin, RD Sands et al. 2014. Climate change effects on agriculture: Economic responses to biophysical shocks. Proc. Natl. Acad. Sci. U. S. A., 111, 3274–3279.

Nelson, GC, MW Rosegrant, J Koo et al. 2009. Climate Change: Impact on Agriculture and Costs of Adaptation. Washington, DC: International Food Policy Research Institute.

Neuhoff, D, J Kwesiga. 2021. Para-organic intensification of future farming as alternative concept to reactor-based staple food production in Africa. Org. Agr., 11, 209–215. https://doi.org/10.1007/s13165-020-00326-y.

Neumann, B, AT Vafeidis, J Zimmermann, RJ Nicholls 2015. Future coastal population growth and exposure to sea-level rise and coastal flooding – A global assessment PLOS One, 10 (3). https://doi.org/10.1371/journal.pone.0118571

Newton, AC, PM Evans, SCL Watson, et al. 2021. Ecological restoration of agricultural land can improve its contribution to economic development. PLOS One, 16(3), e0247850. https://doi.org/10.1371/journal.pone.0247850

Newton, J, CJ Paci, A Ogden 2005. Climate Change and Natural Hazards in Northern Canada: Integrating Indigenous Perspectives with Government Policy. In: Haque, C.E. (eds.) Mitigation of Natural Hazards and Disasters: International Perspectives. Springer, Dordrecht. https://doi.org/10.1007/1-4020-4514-X_11

Ngai, ST, L Juneng, F Tangang et al. 2022. Projected mean and extreme precipitation based on bias-corrected simulation outputs of CORDEX Southeast Asia. Weather Clim. Extrem., 37, 100484.

Nguyen, PL, M Bador, LV Alexander et al. 2023. Selecting regional climate models based on their skill could give more credible precipitation projections over the complex Southeast Asia region. Clim. Dyn. https://doi.org/10.1007/s00382-023-06751-5

Nhamo, L, S Mpandeli, S Liphadzi et al. 2023. Advances in water research: enhancing sustainable water use in irrigated agriculture in South Africa. In Ting, D. S.-K.; O'Brien, P. G. (Eds.), Progress in Sustainable Development: Sustainable Engineering Practices. Amsterdam, Netherlands: Elsevier, pp. 233–248. https://doi.org/10.1016/B978-0-323-99207-7.00007-5

Niang, I, OC Ruppel, MA Abdrabo 2014. Climate Change: Impacts, Adaptation, and Vulnerability. Part B: Regional Aspects. Contribution of Working Group II to the Fifth Assessment Report of the Intergovernmental Panel on Climate Change. Cambridge: Cambridge University Press, pp. 1199–1265.

Nicholls, RJ, PP Wong, VR Burkett et al. 2007. Coastal systems and lowlying areas. In: Parry ML, OF Canziani, JP Palutikof et al. (eds.), Climate Change 2007: Impacts, Adaptation and Vulnerability. Contribution of Working Group II to the Fourth Assessment Report of the Intergovernmental Panel on Climate Change. Cambridge: Cambridge University Press.

Nicholson SE 2017. Climate and climatic variability of rainfall over eastern Africa. Rev. Geophys., 55(3), 590–635. https://doi.org/10.1002/2016RG000544

Nicholson, SE, C Funk, AH Fink 2018. Rainfall over the African continent from the 19th through the 21st century. Glob. Planet. Change, 165, 114–127. https://doi.org/10.1016/j.gloplacha.2017.12.014

Nishanov, N 2015. Sustainable Livestock Management Under Changing Climate in Central Asia. Dushanbe, Tajikistan. https://hdl.handle.net/20.500.11766/4678

Nishioka 2016. Enabling Asia to Stabilise the Climate. Springer Singapore. https://doi.org/10.1007/978-981-287-826-7

Nixon, S 2015. EU Overview of Methodologies Used in Preparation of Flood Hazard and Flood Risk Maps. Brussels: European Commission.

Nkonya, E, A Mirzabaev, J von Braun 2016. Economics of land degradation and improvement: an introduction and overview. In Economics of Land Degradation and Improvement – A Global Assessment for Sustainable Development (pp. 1–14). Cham: Springer.

Nkonya W, E Anderson, J Kato et al. 2016. Global cost of Land degradation. In: E Nkonya, A Mirzabaev, J von Braun (Eds.), Economics of Land Degradation and Improvement – A Global Assessment for Sustainable Development. Cham: Springer International Publishing, pp. 117–165.

NOAA 2022. National Oceanic and Atmospheric Administration. Climate at a glance. www.ncdc.noaa.gov/cag

Nor Diana, MI, NA Zulkepli, C Siwar et al. 2022. Farmers' adaptation strategies to climate change in southeast Asia: A systematic literature review. Sustainability, 14(6), 3639. https://doi.org/10.3390/su14063639

Northrup, DL, B Basso, MQ Wang et al. 2021. Novel technologies for emission reduction complement conservation agriculture to achieve negative emissions from row-crop production. Proc. Natl. Acad. Sci. U. S. A., 118(28), e2022666118. https://doi.org/10.1073/pnas.2022666118

Nouri, A, S Lukas, S Singh et al. 2022. When do cover crops reduce nitrate leaching? A global meta-analysis. Glob. Change Biol., 28(15), 4736–4749. https://doi.org/10.1111/gcb.16269

Ntinyari W, J Gweyi-Onyango, M Giweta et al. 2022. Nitrogen use efficiency trends for sustainable crop productivity in Lake Victoria basin: Smallholder farmers' perspectives on nitrogen cycling. Environ. Res. Commun., 4, 015004. https://doi.org/10.1088/2515-7620/ac40f2

Ntziachristos, L, D Gkatzoflias, C Kouridis et al. 2009. COPERT: A European Road Transport Emission Inventory Model, in: Information Technologies in Environmental Engineering, Environmental Science and Engineering. Berlin, Heidelberg: Springer, pp. 491–504. https://doi.org/10.1007/978-3-540-88351-7_37

Nyadzi, E, E Bessah, G Kranjac-Berisavljevic 2021. Taking stock of climate change induced sea level rise across the West African coast. Environ. Claims J., 33(1), 77–90. https://doi.org/10.1080/10406026.2020.1847873

Nyasimi, M, D Amwata, L Hove et al. 2014. Evidence of impact: climate-smart agriculture in Africa. In: CCAFS working paper no. 86.

O'Neill, BC, C Tebaldi, DP Vuuren et al. 2016. The scenario model intercomparison project (ScenarioMIP) for CMIP6. Geosci. Model Dev., 9, 3461–3482.

Obembe, OS, NP Hendricks, J Tack 2021. Decreased wheat production in the USA from climate change driven by yield losses rather than crop abandonment. PLOS One, 16(6), e0252067. https://doi.org/10.1371/journal.pone.0252067

Oenema, O, F Brentrup, J Lammel et al. 2015. Nitrogen Use Efficiency (NUE) – an Indicator for the Utilization of Nitrogen in Agriculture and Food Systems Wageningen University.

Ofori, SA, SJ Cobbina, S Obiri 2021. Climate change, land, water, and food security: Perspectives from Sub-Saharan Africa. Front. Sustain. Food Syst., 5, 680924. https://doi.org/10.3389/fsufs.2021.680924

Ofosu, E, A Bazrgar, B Coleman et al. 2021. Soil organic carbon enhancement in diverse temperate riparian buffer systems in comparison with adjacent agricultural soils. Agroforest Syst. https://doi.org/10.1007/s10457-021-00691-x

Ogundari, K, R Onyeaghala 2021. The effects of climate change on African agricultural productivity growth revisited. Environ. Sci. Pollut. Res., 28, 30035–30045. https://doi.org/10.1007/s11356-021-12684-5

Okori, P, R Chirwa, V Chisale et al. 2022. New High-Yielding, Stress-Resilient, and Nutritious Crop Varieties. CABI Books. CABI International. https://doi.org/10.1079/9781800621602.0002

Olesen, JE 2016. Socio-economic Impacts – Agricultural Systems. In: Quante, M. & Colijn, F. (eds.) North Sea Region Climate Change Assessment NOSCCA. Springer Nature.

Olesen, JE, M Bindi 2002. Consequences of climate change for European agricultural productivity, land use and policy. Eur. J. Agron., 16(4), 239–262.

Olesen, JE, M Trnka, KC Kersebaum et al. 2011. Impacts and adaptation of European crop production systems to climate change Eur. J. Agron., 34, 96–112, https://doi.org/10.1016/j.eja.2010.11.003

Oliver, V, N Cochrane, J Magnusson et al. 2019. Effects of water management and cultivar on carbon dynamics, plant productivity and biomass allocation in European rice systems. Sci. Total Environ., 685, 1139–1151.

Ologeh, IO, JB Akarakiri, VO Sobanke 2021. Assessment of Climate Change-Induced Migration of Crop Farmers in Northern Nigeria. In: Luetz JM, D Ayal (eds.), Handbook of Climate Change Management. Cham: Springer. https://doi.org/10.1007/978-3-030-57281-5_231

Omollo, E, L Cramer, L Motaroki et al. 2020. Trends and the Future of Livestock Production Systems under a Changing Climate in Africa. AGNES, Policy Brief No. 6, December, 2020.

Ongoma, V, G Tan, BA Ogwang et al. 2015. Diagnosis of seasonal rainfall variability over East Africa: A case study of 2010–2011 drought over Kenya. Pak. J. Meteorol., 11, 13–21.

Ongoma, V, H Chen 2017. Temporal and spatial variability of temperature and precipitation over East Africa from 1951 to 2010. Theor. Appl. Clim., 129, 131–144.

Ongoma, V, H Chena, C Gaoa 2018. Projected changes in mean rainfall and temperature over east Africa based on CMIP5 models. Int. J. Climatol., https://doi.org/10.1002/joc.5252

Ongoma, V, T Epule, Y Brouziyne et al. 2023. COVID-19 response in Africa: Impacts and lessons for environmental management and climate change adaptation. Environ. Dev. Sustain. https://doi.org/10.1007/s10668-023-02956-0

Onyutha, C 2021. Trends and variability of temperature and evaporation over the African continent: Relationships with precipitation. Atmósfera, 34(3), 267–287. https://doi.org/10.20937/ATM.52788

Ort, DR, SP Long 2014. Limits on yields in the Corn Belt. Science, 344, 484–485.

Ortas, I, R Lal 2013. Food Security and Climate Change in West Asia. In: Sivakumar, M., Lal, R., Selvaraju, R., Hamdan, I. (eds.), Climate Change and Food Security in West Asia and North Africa. Dordrecht: Springer. https://doi.org/10.1007/978-94-007-6751-5_12

Ortiz, R, KD Sayre, B Govaerts et al. 2008. Climate change: Can wheat beat the heat? Agric. Ecosyst. Environ., 126, 46–58. https://doi.org/10.1016/j.agee.2008.01.019

Osman-Elasha, B, DF de Velasco 2021. Drivers of Greenhouse Gas Emissions in Africa: Focus on Agriculture, Forestry and Other Land Use. June 22 African Development Bank (2021) Retrieved from https://blogs.afdb.org/climate-change-africa/drivers-greenhouse-gas-emissions-africa-focus-agriculture-forestry-and-other

Otte, J, U Pica-Ciamarra, S Morzaria 2019. A comparative overview of the livestock-environment interactions in Asia and Sub-Saharan Africa. Front. Vet. Sci., 6, 37. https://doi.org/10.3389/fvets.2019.00037

Ouédraogo, S, M Bawindsom Kébré, F Zougmoré 2021. water dynamics under drip irrigation to proper manage water use in arid zone. J. Agric. Chem. Environ., 10, 57–68. https://doi.org/10.4236/jacen.2021.101004

Oumara, NGA, L El Youssfi 2022. Salinization of soils and aquifers in Morocco and the alternatives of response. Environ. Sci. Proc., 16, 65. https://doi.org/10.3390/environsciproc2022016065

Ouyang, Z, RB Jackson, G McNicol et al. 2023. Paddy rice methane emissions across Monsoon Asia. Remote Sens. Environ., 284, 113335. https://doi.org/10.1016/j.rse.2022.113335

Overpeck, JT, B Udall 2020. Climate change and the aridification of North America. Proc. Natl. Acad. Sci. U. S. A., 117, 11856–11858.

Ozer, P, A Mahamoud 2013. Recent extreme precipitation and temperature changes in Djibouti City (1966–2011). J. Climatol., 2013, 928501.

Özkan, S, A Vitali, N Lacetera et al. 2016. Challenges and priorities for modelling livestock health and pathogens in the context of climate change. Environ. Res., 151, 130–144. https://doi.org/10.1016/j.envres.2016.07.033

Page, ER, S Meloche, M Moran et al. 2021. Effect of seeding date on winter canola (*Brassica napus* L.) yield and oil quality in southern Ontario. Can. J. Plant Sci., 101(4), 490–499. https://doi.org/10.1139/cjps-2020-0220

Pal, I, BT Anderson, GD Salvucci et al. 2013. Shifting seasonality and increasing frequency of precipitation in wet and dry seasons across the US. Geophys. Res. Lett., 40, 4030–4035.

Pan, S, X Zhao, Y Han et al. 2022. Online information platform for the management of national variety test of major crops in China: Design, development, and applications. Comput. Electron. Agric, 201, 107292.

Panagos, P, A Imeson, K Meusburger et al. 2016. Soil conservation in Europe: Wish or reality? Land Degrad. Dev., 27(6), 1547–1551.

Panagos, P, C Ballabio, M Himics et al. 2021. Projections of soil loss by water erosion in Europe by 2050. Environ. Sci. Policy, 124, 380–392. https://doi.org/10.1016/j.envsci.2021.07.012

Panagos, P, G Standardi, P Borrelli et al. 2018. Cost of agricultural productivity loss due to soil erosion in the European Union: From direct cost evaluation approaches to the use of macroeconomic models. Land Degrad. Dev., 29, 471–484.

Papalexiou, SM, A Montanari 2019. Global and regional increase of precipitation extremes under global warming. Water Res. Res., 55, 4901–4914. https://doi.org/10.1029/2018WR024067

Paraschivu, M, A Olaru 2020. Effects of interaction between abiotic stress and pathogens in cereals in the context of climate change: An overview. Ann. Univ. Craiova Agric. Montanol. Cadastre Ser., 49(2), 413–424.

Park, S, J Lee, J Yeom et al. 2022. Performance of drought indices in assessing rice yield in North Korea and South Korea under the different agricultural systems. Remote Sens., 14, 6161. https://doi.org/10.3390/rs14236161

Pasley, H, H Brown, D Holzworth et al. 2023. How to build a crop model. A review. Agron. Sustain. Dev., 43, 2. https://doi.org/10.1007/s13593-022-00854-9

Pasquel, D, S Roux, J Richetti et al. 2022. A review of methods to evaluate crop model performance at multiple and changing spatial scales. Precision Agric., 23, 1489–1513. https://doi.org/10.1007/s11119-022-09885-4

Pasqui, M, ED Giuseppe 2019. Climate change, future warming and adaptation in Europe. Anim. Front., 9(1), 6–11.

Pathak TB, ML Maskey, JA Dahlberg et al. 2018. Climate change trends and impacts on California agriculture: A detailed review. Agronomy, 8, 25. https://doi.org/10.3390/agronomy8030025

Patrick, E 2017. Drought characteristics and management in Central Asia and Turkey. In FAO Water Reports; Food and Agriculture Organization of the United Nations Rome: Rome, Italy, 2017; Available online at: http://www.fao.org/3/a-i6738e.pdf

Paudel, S, H Baral, A Rojario et al. 2022. Agroforestry: Opportunities and challenges in Timor-Leste. Forests, 13, 41.

Peltonen-Sainio, P, L Jauhiainen 2020. Large zonal and temporal shifts in crops and cultivars coincide with warmer growing seasons in Finland. Reg. Environ. Change, 20, 89. https://doi.org/10.1007/s10113-020-01682-x.

Pereira, L 2017. Climate Change Impacts on Agriculture across Africa. In: Oxford Research Encyclopedia of Environmental Science. Oxford, UK: Oxford University Press. https://doi.org/10.1093/acrefore/9780199389414.013.292

Picketts, IM, AT Werner, TQ Murdock 2009. Climate Change in Prince George: Summary of Past Trends and Future Projections; Pacific Climate Impacts Consortium, University of Victoria, Victoria BC, 48 p.

Pickson, RB, P Gui, A Chen et al. 2023. Climate change and food security nexus in Asia: A regional comparison. Ecol. Inform., 76, 102038. https://doi.org/10.1016/j.ecoinf.2023.102038

Pimentel, D, M Burgess 2013. Soil erosion threatens food production. Agriculture, 3, 443–463. https://doi.org/10.3390/agriculture3030443

Pimentel, D, M Burgess 2014. Biofuel production using food. Environ. Dev. Sustain., 16, 1–3. https://doi.org/10.1007/s10668-013-9505-6.

Pineda, PS, EB Flores, JRV Herrera et al. 2021. Opportunities and challenges for improving the productivity of swamp buffaloes in Southeastern Asia. Front. Genet., 12:629861. https://doi.org/10.3389/fgene.2021.629861

Plötz, P, J Wachsmuth, T Gnann et al. 2021. Net-Zero-Carbon Transport in Europe Until 2050 – Targets, Technologies and Policies for a Long-Term EU Strategy. Karlsruhe: Fraunhofer Institute for Systems and Innovation Research ISI.

Poesen, J 2018. Soil erosion in the Anthropocene: Research needs. Earth Surf. Process. Landf, 43, 64–84. https://doi.org/10.1002/esp.4250

Popp, A, K Calvin, S Fujimori et al. 2017. Land-use futures in the shared socio-economic pathways. Glob. Environ. Change, 42, 331–345.

Portmann, FT, S Siebert, P Döll 2010. MIRCA2000; Global monthly irrigated and rainfed crop areas around the year 2000: A new high-resolution data set for agricultural and hydrological modeling. Glob. Biogeochem. Cycles, 24, GB1011.

Portner, HO, DC Roberts, H Adams et al. 2022. Climate Change: Impacts, Adaptation and Vulnerability. Intergovernmental Panel on Climate Change (2022).

Pradel, W, M Gatto, G Hareau et al. 2019. Adoption of potato varieties and their role for climate change adaptation in India. Clim. Risk Manage., 23, 114–123. https://doi.org/10.1016/j.crm.2019.01.001

Prein, AF, A Gobiet, H Truhetz et al. 2015. Precipitation in the EURO-CORDEX 0.11∘ simulations: high resolution, high benefits? Clim. Dyn, 46(1-2), 383–412.

PRIMAP 2016. Factsheets: Intended nationally determined contributions (INDCs). Retrieved from https://www.pik-potsdam.de/primaplive/indc-factsheets/

Puertas, R, L Marti, C Calafat 2023. Agricultural and innovation policies aimed at mitigating climate change. Environ. Sci. Pollut. Res. https://doi.org/10.1007/s11356-023-25663-9

Punyawaew, K, D Suriya-arunroj, M Siangliw et al. 2016. Thai jasmine rice cultivar KDML105 carrying Saltol QTL exhibiting salinity tolerance at seedling stage. Mol. Breed., 36, 150.

Qian, B, Q Jing, G Bélanger et al. 2018. Simulated canola yield responses to climate change and adaptation in Canada. Agron. J., 110, 133–46.

Qian, B, X Zhang, W Smith et al. 2019. Climate change impacts on Canadian yields of spring wheat, canola and maize for global warming levels of 1.5°C, 2.0°C, 2.5°C and 3.0°C. Environ. Res. Lett., 14, 074005.

Quemada, M, L Lassaletta, LS Jensen et al. 2020. Exploring nitrogen indicators of farm performance among farm types across several European case studies. Agric. Syst., 177, 102689.

Quenum, GMLD, F Nkrumah, NAB Klutse et al. 2021. Spatiotemporal changes in temperature and precipitation in West Africa. Part I: Analysis with the CMIP6 historical dataset. Water, 13, 3506. https://doi.org/10.3390/w13243506

Radić, V, A Bliss, AC Beedlow et al. 2014. Regional and global projections of twenty-first century glacier mass changes in response to climate scenarios from global climate models. Clim. Dyn., 42, 37–58. https://doi.org/10.1007/s00382-013-1719-7

Rafik, A, H Ibouh, A El Alaoui El Fels et al. 2022. Soil salinity detection and mapping in an environment under water stress between 1984 and 2018 (Case of the largest oasis in Africa-Morocco). Remote Sens., 14, 1606. https://doi.org/10.3390/rs14071606

Rahimi, J, E Fillol, JY Mutua et al. 2022. A shift from cattle to camel and goat farming can sustain milk production with lower inputs and emissions in north sub-Saharan Africa's drylands. Nat. Food, 3, 523–531. https://doi.org/10.1038/s43016-022-00543-6

Rahman, MZ 2013. The relationship between trade openness and carbon emission: A case of Bangladesh. J. Empir. Econ., 1(4), 126–134.

Rahman, NMF, WA Malik, MS Kabir et al. 2023. 50 years of rice breeding in Bangladesh: Genetic yield trends. Theor. Appl. Genet., 136(1), 18. https://doi.org/10.1007/s00122-023-04260-x

Rajczak, J, S Kotlarski, N Salzmann et al. 2016. Robust climate scenarios for sites with sparse observations: A two-step bias correction approach. Int. J. Climatol., 36, 1226–1243.

Ramseur, JL 2014. U.S. Greenhouse Gas Emissions: Recent Trends and Factors. New York, USA: Congressional Research Service.

Ray, DK, ND Mueller, PC West et al. 2013. Yield trends are insufficient to double global crop production by 2050. PLOS One, 8(6), e66428. https://doi.org/10.1371/journal.pone.0066428

Ray, DK, PC West, M Clark et al. 2019. Climate change has likely already affected global food production. PLOS One, 14, e0217148.

Ray, RL, VP Singh, SK Singh et al. 2022. What is the impact of COVID-19 pandemic on global carbon emissions? Sci. Total Environ., 816, 151503.

Raymundo, R, S Asseng, R Robertson et al. 2018. Climate change impact on global potato production. Eur. J. Agron., 100, 87–98.

Razzaq, A, P Kaur, N Akhter et al. 2021. Next-generation breeding strategies for climate-ready crops. Front. Plant Sci., 12, 620420. https://doi.org/10.3389/fpls.2021.620420

RBC 2022. The Next Green Revolution: How Canada can produce more food and fewer emissions. RBC, Canada. https://thoughtleadership.rbc.com/the-next-green-revolution-how-canada-can-produce-more-food-and-fewer-emissions/

Reddy, SK, S Liu, JC Rudd et al. 2014. Physiology and transcriptomics of water-deficit stress responses in wheat cultivars TAM 111 and TAM 112. J. Plant Physiol., 171(14), 1289–1298.

Rees, RM, J Maire, A Florence et al. 2020. Mitigating nitrous oxide emissions from agricultural soils by precision management. Front. Agric. Sci. Eng., 7(1), 75–80.

Reinders, FB 2011. Irrigation methods for efficient water application: 40 years of South African research excellence. Water SA, 37(5), 765–770.

Ren, G, JCL Chan, H Kubota et al. 2021. Historical and recent change in extreme climate over East Asia. Clim. Change, 168(3–4), 22. https://doi.org/10.1007/s10584-021-03227-5

Ren, X, D Sun, Q Wang 2016. Modeling the effects of plant density on maize productivity and water balance in the Loess Plateau of China. Agric. Water Manag., 171, 40–48.

Restovich, SB, AE Andriulo, SI Portela 2022. Cover crop mixtures increase ecosystem multifunctionality in summer crop rotations with low N fertilization. Agron. Sustain. Dev., 42, 19. https://doi.org/10.1007/s13593-021-00750-8

Reyer, CPO, IM Otto, S Adams et al. 2017. Climate change impacts in Central Asia and their implications for development. Reg. Environ. Change, 17(6), 1639–1650.

Rhodes, EE, A Jalloh, A Diouf 2014. Review of research and policies for climate change adaptation in the agriculture sector in West Africa. Future Agricultures Working Paper 90.

Ringeval, B, C Müller, TAM Pugh et al. (2021) Potential yield simulated by global gridded crop models: Using a process-based emulator to explain their differences. Geosci. Model Dev., 14, 1639–1656. https://doi.org/10.5194/gmd-14-1639-2021

Rippke, U, J Ramirez-Villegas, A Jarvis et al. 2016. Timescales of transformational climate change adaptation in sub-Saharan African agriculture. Nat. Clim. Change, 6(6), 605–609.

Ritchie, H, M Roser 2020. Emissions by sector. Our World Data. https://ourworldindata.org/emissions-by-sector.

Ritchie, H, M Roser, P Rosado 2020. CO_2 and Greenhouse Gas Emissions. Published online at OurWorldInData.org. Retrieved from: https://ourworldindata.org/co2-and-greenhouse-gas-emissions [Online Resource].

Ritchie, H, P Rosado, M Roser 2017. Meat and Dairy Production. Published online at OurWorldInData.org. Retrieved from: https://ourworldindata.org/meat-production [Online Resource].

Robertson, GP, TW Bruulsema, RJ Gehl et al. 2013. Nitrogen–climate interactions in US agriculture. Biogeochemistry, 114, 41–70. https://doi.org/10.1007/s10533-012-9802-4

Robine, JM, SLK Cheung, S Le Roy et al. 2008. Death toll exceeded 70 000 in Europe during the summer of 2003. C.R. Biol., 331, 171–178.

Rocha, M, M Krapp, J Guetschow et al. 2015. Historical responsibility for climate change— From countries emissions to contribution to temperature increase. Climate Analytics https://climateanalytics.org/media/historical_responsibility_report_nov_2015.pdf

Rockström, J, M Falkenmark 2015. Agriculture: Increase water harvesting in Africa. Nature, 519, 283–285.

Rodale Institute 2012. The Farming Systems Trial: Celebrating 30 Years. Pennsylvania: Rodale Press.

Roesch-McNally, GE, J Gordon Arbuckle, JC Tyndall 2017. What would farmers do? Adaptation intentions under a Corn Belt climate change scenario. Agric. Hum. Values, 34, 333–346. https://doi.org/10.1007/s10460-016-9719-y

Rojas-Downing, MM, AP Nejadhashemi, T Harrigan et al. 2017. Climate change and livestock: Impacts, adaptation, and mitigation Clim Risk Manag, 16, 145–163. https://doi.org/10.1016/j.crm.2017.02.001

Rosa, L 2022. Adapting agriculture to climate change via sustainable irrigation: Biophysical potentials and feedbacks. Environ. Res. Lett., 17, 063008. https://doi.org/10.1088/1748-9326/ac7408

Rosenstock, TS, IK Dawson, E Aynekulu et al. 2019. A planetary health perspective on agroforestry in sub-Saharan Africa. One Earth, 1, 330–344. https://doi.org/10.1016/j.oneear.2019.10.017

Rosenzweig C, A Iglesias, XB Yang et al. 2001. Climate change and extreme weather events – implications for food production, plant diseases, and pests. Glob. Change Hum. Health, 2, 90–104. https://doi.org/10.1023/A:1015086831467

Rosenzweig, C, J Elliott, D Deryng et al. 2014. Assessing agricultural risks of climate change in the 21st century in a global gridded crop model intercomparison. Proc. Natl. Acad. Sci. U. S. A., 111, 3268–3273.

Röttgers, D, B Anderson 2018. Power Struggle: Decarbonising the Electricity Sector Research Papers in Economics. Paris: OECD Publishing. https://doi.org/10.1787/900f4c72-en

Roudier, P, B Muller, P D'Aquino et al. 2014. The role of climate forecasts in smallholder agriculture: Lessons from participatory research in two communities in Senegal. Clim. Risk Manag., 2, 42–55. https://doi.org/10.1016/j.crm.2014.02.001

Roudier, P, S Sultan, P Quirion et al. 2011. The impact of future climate change on West African crop yields: What does the recent literature say? Glob. Environ. Change 21, 1073–1083. https://doi.org/10.1016/j.gloenvcha.2011.04.007

Ruosteenoja K, J Räisänen, A Venäläinen et al. 2016. Projections for the duration and degree days of the thermal growing season in Europe derived from CMIP5 model output. Int. J. Climatol., 36, 3039–3055. https://doi.org/10.1002/joc.4535

Ruosteenoja, K, T Markkanen, J Raisanen 2020. Thermal season in northern Europe in projected future climate. Int. J. Climatol., 40, 4444–4462. https://doi.org/10.1002/joc.6466

Russo, S, J Sillmann, EM Fischer 2015. Top ten European heatwaves since 1950 and their occurrence in the coming decades. Environ. Res. Lett., 10, 124003.

Saddique, N, M Jehanzaib, A Sarwar et al. 2022. Systematic review on Farmers' adaptation strategies in Pakistan toward climate change. Atmosphere, 13, 1280. https://doi.org/10.3390/atmos13081280

Sakadevan, K, M Nguyen 2010. Extent, impact, and response to soil and water salinity in arid and semiarid regions. Adv. Agron., 109, 55–74. https://doi.org/10.1016/B978-0-12-385040-9.00002-5

Salack, S, S Sanfo, M Sidibe et al. 2022. Low-cost adaptation options to support green growth in agriculture, water resources, and coastal zones. Sci. Rep., 12, 17898. https://doi.org/10.1038/s41598-022-22331-9

Salahuddin, M, J Gow, I Ali et al. 2019. Urbanization-globalization-CO_2 emissions nexus revisited: Empirical evidence from South Africa. Heliyon, 5(6), e01974. https://doi.org/10.1016/j.heliyon.2019.e01974

Salciccioli, JD, Y Crutain, M Komorowski et al. 2016. Sensitivity Analysis and Model Validation. In: Secondary Analysis of Electronic Health Records. Cham: Springer. https://doi.org/10.1007/978-3-319-43742-2_17

Saleem, M 2022. Possibility of utilizing agriculture biomass as a renewable and sustainable future energy source. Heliyon, 8(2), e08905.

Saloux, E, JA Candanedo 2018. Forecasting district heating demand using machine learning algorithms. Energy Proc., 149, 59–68.

Salumbo, A 2020. A review of soil erosion estimation methods. Agric. Sci., 11, 667–691. https://doi.org/10.4236/as.2020.118043

Sanderson, EW, J Walston, JG Robinson 2018. From bottleneck to breakthrough: Urbanization and the future of biodiversity conservation. BioScience, 68, 412–426.

Sanjay, J, JV Revadekar, MVS RamaRao et al. 2020. Temperature Changes in India. Assessment of Climate Change Over the Indian Region. Singapore: Springer, pp. 21–45.

Santini, M, S Noce, M Antonelli et al. 2022. Complex drought patterns robustly explain global yield loss for major crops. Sci. Rep., 12, 5792.

Santos 2017. Road transport and CO_2 emissions: What are the challenges? Transp. Policy, 59, 71–74.

Sarkar, SF, SP Jacquelyne, L Etienne et al. 2018. Enabling a sustainable and prosperous future through science and innovation in the bioeconomy at Agriculture and Agri-Food Canada. New Biotechnol., 40(Pt A), 70–75. https://doi.org/10.1016/j.nbt.2017.04.001

Sarker, MAR, K Alam, J Gow 2012. Exploring the relationship between climate change and rice yield in Bangladesh: An analysis of time series data. Agric. Syst., 112, 11–16.

Sarr, C, M Ndour, M Haddad et al. 2021. Estimation of sea level rise on the West African coasts: Case of Senegal, Mauritania and Cape Verde. Int. J. Geosci., 12, 121–137. https://doi.org/10.4236/ijg.2021.122008

Sartori, M, G Philippidis, E Ferrari et al. 2019. A linkage between the biophysical and the economic assessing the global market impacts of soil erosion. Land Use Policy, 86, 299–312.

Saturday, A 2018. Restoration of degraded agricultural land: A review. J. Environ. Health Sci., 4(2), 44–51. https://www.ommegaonline.org/article-details/Restoration-of-Degraded-Agricultural-Land-A-Review/1928

Sauchyn, D, S Kulshreshtha 2010. Agriculture adaptation through irrigation. In: Sauchyn D, Diaz P, Kulshreshtha SN, editors. The New Normal: The Canadian Prairies in a Changing Climate. Regina: Canadian Plains Research Centre, University of Regina.

Saud, S, D Wang, S Fahad et al. 2022. Comprehensive impacts of climate change on rice production and adaptive strategies in China. Front. Microbiol., 13, 926059. https://doi.org/10.3389/fmicb.2022.926059

Scarlat, N, M Prussi, M Padella 2022. Quantification of the carbon intensity of electricity produced and used in Europe. Appl. Energy, 305, 117901.

Scavo, A, S Fontanazza, A Restuccia et al. 2022. The role of cover crops in improving soil fertility and plant nutritional status in temperate climates. A review. Agron. Sustain. Dev., 42, 93. https://doi.org/10.1007/s13593-022-00825-0

Schauberger, B, S Archontoulis, A Arneth et al. 2017. Consistent negative response of US crops to high temperatures in observations and crop models. Nat. Commun., 8, 13931. https://doi.org/10.1038/ncomms13931

Scherhaufer, S, G Moates, H Hartikainen et al. 2018. Environmental impacts of food waste in Europe. Waste Manag., 77, 98–113.

Schilling, J, E Hertig, Y Tramblay et al. 2020. Climate change vulnerability, water resources and social implications in North Africa. Reg. Environ. Change, 20, 15. https://doi.org/10.1007/s10113-020-01597-7

Schmitt, J, F Offermann, M Soder et al. 2022. Extreme weather events cause significant crop yield losses at the farm level in German agriculture. Food Policy, 112, 102359.

Schnell, R, K Horn, E Biar et al. 2021. Texas corn performance variety trials. Texas A&M AgriLife Research and AgriLife Extension Service. Retrieved from: https://varietytesting.tamu.edu.

Schnitkey, GD, SC Sellars, LF Gentry 2023. Cover Crops, Farm Economics, and Policy. Paper Presented at Allied Social Science Association Meetings, January 6–8, 2023.

Searchinger, T, M Herrero, X Yan et al. 2021. Opportunities to Reduce Methane Emissions from Global Agriculture. Available at: https://searchinger.princeton.edu/document/46

Searchinger, T, R Waite, C Hanson et al. 2019. Creating a Sustainable Food Future: A Menu of Solutions to Feed Nearly 10 Billion People by 2050. Final report. Washington, DC: World Resources Institute. https://research.wri.org/sites/default/files/2019-07/WRR_Food_Full_Report_0.pd

Seidel, SJ, T Gaiser, AK Srivastava et al. 2022. Simulating root growth as a function of soil strength and yield with a field-scale crop model coupled with a 3D architectural root model. Front. Plant Sci., 13, 865188. https://doi.org/10.3389/fpls.2022.865188

Sengar, RS, K Sengar 2014. Climate Change Effect on Crop Productivity. London: CRC Press.

Serdeczny, O, S Adams, F Baarsch et al. 2017. Climate change impacts in Sub-Saharan Africa: From physical changes to their social repercussions. Reg. Environ. Change, 17, 1585–600.

Setimela, PS, C Magorokosho, R Lunduka et al. 2017. On-farm yield gains with stress tolerant maize in Eastern and Southern Africa. Agron. J., 109, 406–417.

Shah, SMH, Z Mustaffa, FY Teo et al. 2020. A review of the flood hazard and risk management in the South Asian Region, particularly Pakistan. Sci. Afr., 00651. https://doi.org/10.1016/j.sciaf.2020.e00651

Shahariar, S, D Peak, R Soolanayakanahally et al. 2021. Impact of short-rotation willow as riparian land-use practice on soil organic carbon fractions and composition from two contiguous wetland systems in the prairie pothole region. Agroforest Syst. https://doi.org/10.1007/s10457-021-00694-8

Shahid, M, KU Rahman 2021. Identifying the annual and seasonal trends of hydrological and climatic variables in the Indus Basin Pakistan. Asia-Pac. J. Atmos. Sci., 57, 191–205.

Shahzad, A, S Ullah, AF Dar et al. 2021. Nexus on climate change: Agriculture and possible solution to cope future climate change stresses. Environ. Sci. Pollut. Res., 28(12), 14211–14232.

Shang, Z, M Abdalla, L Xia et al. 2021. Can cropland management practices lower net greenhouse emissions without compromising yield? Glob. Change Biol., 27, 4657–4670.

Sharma, G, LK Sharma 2022. Climate change effect on soil carbon stock in different land use types in eastern Rajasthan, India. Environ. Dev. Sustain., 24, 4942–4962. https://doi.org/10.1007/s10668-021-01641-4

Sharma, RK, S Kumar, K Vatta et al. 2022. Impact of recent climate change on corn, rice, and wheat in southeastern USA. Sci. Rep., 12, 16928. https://doi.org/10.1038/s41598-022-21454-3

Sheffield, J, EF Wood, N Chaney et al. 2014. A drought monitoring and forecasting system for sub-Sahara African water resources and food security. Bull. Am. Meteor. Soc., 95, 861–882. https://doi.org/10.1175/BAMS-D-12-00124.1

Sheikh, MM, N Manzoor, M Adnan et al. 2009. Climate Profile and Past Climate Changes in Pakistan; Technical report. Islamabad, Pakistan: Global Change Impact Studies Centre.

Shem Juma, G, F Kelonye Beru 2021. Prediction of Crop Yields under a Changing Climate. IntechOpen. https://doi.org/10.5772/intechopen.94261

Shen, Y, C Jiang, KL Chan et al. 2021. Estimation of field-level Nox emissions from crop residue burning using remote sensing data: A case study in Hubei, China. Remote Sens., 13(3), 404. https://doi.org/10.3390/rs13030404

Shikwambana, S, N Malaza, K Shale 2021. Impacts of rainfall and temperature changes on smallholder agriculture in the Limpopo Province, South Africa. Water, 13(20), 2872. https://doi.org/10.3390/w13202872

Shilenje ZW, V Ongoma, M Njagi 2019. Applicability of combined drought index in drought analysis over North Eastern Kenya. Nat. Hazards, 99, 379–389. https://doi.org/10.1007/s11069-019-03745-7

Shock, CC, BM Shock, T Welch 2013. Strategies for efficient irrigation water use. EM 8783. Oregon State University, Department of Crop and Soil Science, Extension Service. Retrieved from EM 8782. http://extension.oregonstate.edu/catalog

Shongwe, ME, GJ Van Oldenborgh, B van den Hurk et al. 2011. Projected changes in mean and extreme precipitation in Africa under global warming. Part II: East Africa. J. Clim., 24, 3718–3733.

Shultz, JM, J Russell, Z Espinel 2005. Epidemiology of tropical cyclones: The dynamics of disaster, disease, and development. Epidemiol. Rev., 27, 21–35.

Si, Z, A Qin, Y Liang et al. 2023. A review on regulation of irrigation management on wheat physiology, grain yield, and quality. Plants (Basel), 12(4), 692. https://doi.org/10.3390/plants12040692

Siarudin, M, SA Rahman, Y Artati et al. 2021. Carbon sequestration potential of agroforestry systems in degraded landscapes in West Java, Indonesia. Forests, 12, 714.

Simon, K 2022. Evaluations of Climate Change Impacts on Crop Yields in Saskatchewan and Ontario and of Opportunities for Agricultural Expansion in Northern Ontario. School of Environmental Sciences, The University of Guelph. https://hdl.handle.net/10214/26977

Simtowe, F, E Amondo, P Marenya et al. 2019. Impacts of drought-tolerant maize varieties on productivity, risk, and resource use: Evidence from Uganda. Land Use Policy, 88, 104091.

Singh, H, MR Najafi, A Cannon 2022. Evaluation and joint projection of temperature and precipitation extremes across Canada based on hierarchical Bayesian modelling and large ensembles of regional climate simulations. Weather Clim. Extrem., 36, 100443.

Sivakumar, MVK, J Hansen 2007. Climate Prediction and Agriculture: Summary and the Way Forward. In: Sivakumar MVK, J Hansen (eds.), Climate Prediction and Agriculture. Berlin, Heidelberg: Springer. https://doi.org/10.1007/978-3-540-44650-7_1

Sivakumar, MVK, R Lal, L Selvaraju et al. 2013. Climate Change and Food Security in West Asia and North Africa. Dordrecht: Springer. https://doi.org/10.1007/978-94-007-6751-5

Sivakumar, MVK, R Stefanski 2011. Climate Change in South Asia. In Lal R, MVK Sivakumar, SMA Faiz et al. (Eds.), Climate Change and Food Security in South Asia. New York and London: Springer, pp. 13–30.

Skarbø, K, K VanderMolen 2016. Maize migration: Key crop expands to higher altitudes under climate change in the Andes. Clim. Dev., 8, 245–255. https://doi.org/10.1080/17565529.2015.1034234

Skobelev, PO, EV Simonova, SV Smirnov et al. 2019. Development of a knowledge base in the "smart farming" system for agricultural enterprise management. Procedia Comp. Sci., 150, 154–161. https://doi.org/10.1016/j.procs.2019.02.029

Sloat, LL, SJ Davis, JS Gerber et al. 2020. Climate adaptation by crop migration. Nat. Commun., 11, 1243. https://doi.org/10.1038/s41467-020-15076-4

Smartt, AD, KR Brye, CW Rogers et al. 2016. Previous crop and cultivar effects on methane emissions from drill-seeded, delayed-flood rice grown on a clay soil. Appl. Environ. Soil Sci. https://doi.org/10.1155/2016/9542361

Smit, B, M Skinner 2002. Adaptation options in agriculture to climate change: A typology. Mitig. Adapt. Strateg. Glob. Change, 7, 85–114.

Smit, R, R Smokers, E Rabé 2007. A new modelling approach for road traffic emissions: VERSIT+. Transp. Res. Part Transp. Environ., 12, 414–422.

Smith, AB, RW Katz 2013. US billion-dollar weather and climate disasters: Data sources, trends, accuracy and biases. Nat. Hazards, 67, 387–410.

Smith, LG, S Westaway, S Mullender et al. 2022. Assessing the multidimensional elements of sustainability in European agroforestry systems. Agric. Syst., 197, 103357. https://doi.org/10.1016/j.agsy.2021.103357

Smith, MM, G Bentrup, T Kellerman et al. 2022. Silvopasture in the USA: A systematic review of natural resource professional and producer-reported benefits, challenges, and management activities. Agric. Ecosyst. Environ., 326, 107818.

Soares, D, TA Paço, J Rolim 2023. Assessing climate change impacts on irrigation water requirements under Mediterranean conditions—A review of the methodological approaches focusing on maize crop. Agronomy, 13, 117. https://doi.org/10.3390/agronomy13010117

Soares JR, BR Souza, AM Mazzetto et al. 2023. Mitigation of nitrous oxide emissions in grazing systems through nitrification inhibitors: A meta-analysis. Nutr. Cycl. Agroecosystems, 125, 359–377. https://doi.org/10.1007/s10705-022-10256-8

Sofer, K 2017. The slow motion crisis: the destabilizing effects of climate change in Turkey and Iraq through 2050. New America.

Sollen-Norrlin, M, BB Ghaley, NLJ Rintoul 2020. Agroforestry benefits and challenges for adoption in Europe and beyond. Sustainability, 12, 7001.

Soul, W, PD Muchaurawa, WD Dorcas 2022. Drought coping strategies by smallholder cattle farmers in Zimbabwe. Trop. Subtrop. Agroecosystems, 25, #069.

Soussana, JF, AI Graux, FN Tubiello 2010. Improving the use of modelling for projections of climate change impacts on crops and pastures. J. Exp. Bot., 61, 2217–2228.

Spinoni, J, J Vogt, P Barbosa 2015. European degree-day climatologies and trends for the period 1951–2011. Int. J. Climatol., 35, 25–36.

Spinoni, J, JV Vogt, G Naumann et al. 2018. Will drought events become more frequent and severe in Europe? Int. J. Climatol., 38, 1718–1736. https://doi.org/10.1002/joc.5291

Springmann, M, D Mason-D'Croz, S Robinson et al. 2016. Global and regional health effects of future food production under climate change: A modelling study. Lancet, 387(10031), 1937–1946.

Stahl, K, I Kohn, V Blauhut et al. 2016. Impacts of European drought events: Insights from an international database of text-based reports, Nat. Hazards Earth Syst. Sci., 16, 801–819, https://doi.org/10.5194/nhess-16-801-2016

Steinfeld, H, T Wassenaar 2007. The role of livestock production on carbon and nitrogen cycles. Annu. Rev. Environ. Resour., 32, 271–294. https://doi.org/10.1146/annurev.energy. 32.041806.143508

Stern, DI, PW Gething, CW Kabaria et al. 2011. Temperature and malaria trends in highland East Africa. PLOS One, 6(9), e24524. https://doi.org/10.1371/journal.pone.0024524

Stetter, C, J Sauer 2022. Greenhouse gas emissions and eco-performance at farm level: A parametric approach. Environ. Resour. Econ., 81, 617–647. https://doi.org/10.1007/s10640-021-00642-1

Stocker, TF, D Qin, GK Plattner et al. 2013. Climate Change 2013 the Physical Science Basis: Working Group I Contribution to the Fifth Assessment Report of the Intergovernmental Panel on Climate Change.

Stoll-Kleemann, S, UJ Schmidt 2017. Schmidt reducing meat consumption in developed and transition countries to counter climate change and biodiversity loss: A review of influence factors. Reg. Environ. Change, 17(5), 1261–1277. https://doi.org/10.1007/s10113-016-1057-5

Stone, P 2001. The effects of heat stress on cereal yield and quality. Pp. 243–291 in A. S. Basra, ed. Crop Responses and Adaptations to Temperature Stress. Binghamton, NY: Food Products Press.

Stroosnijder, L, D Moore, A Alharbi et al. 2012. Improving water use efficiency in drylands. Curr. Opin. Environ. Sustain., 4(5), 497–506.

Stuch, B, J Alcamo, R Schaldach 2021. Projected climate change impacts on mean and year-to-year variability of yield of key smallholder crops in Sub-Saharan Africa. Clim. Dev., 13(3), 1–15.

Su, Y, B Gabrielle, D Makowski 2021. The impact of climate change on the productivity of conservation agriculture. Nat. Clim. Change, 11(7), 628–633.

Sugasti, L, R Pinzón 2020. First approach of abiotic drivers of soil CO_2 efflux in Barro Colorado Island, Panama. Air Soil Water Res., 13, 1178622120960096.

Sultan, B, D Defrance, T Iizumi 2019. Evidence of crop production losses in West Africa due to historical global warming in two crop models. Sci. Rep., 9, 12834. https://doi.org/10.1038/s41598-019-49167-0

Sultan, B, M Gaetani 2016. Agriculture in West Africa in the twenty-first century: Climate change and impacts scenarios, and potential for adaptation. Front. Plant Sci., 7, 1262.

Summer, A, I Lora, P Formaggioni et al. 2018. Impact of heat stress on milk and meat production. Anim. Front., 9(1), 39–46. https://doi.org/10.1093/af/vfy026

Sun, H, X Zhang, E Wang et al. 2016. Assessing the contribution of weather and management to the annual yield variation of summer maize using APSIM in the North China Plain. Field Crop. Res., 194, S0378429016301526.

Sun, J, YP Li, C Suo et al. 2019. Impacts of irrigation efficiency on agricultural water-land nexus system management under multiple uncertainties—A case study in Amu Darya River basin, Central Asia. Agric. Water Manag., 216, 76–88. https://doi.org/10.1016/j.agwat.2019.01.025

Supari, F Tangang, L Juneng et al. 2020. Multi-model projections of precipitation extremes in Southeast Asia based on CORDEX-Southeast Asia simulations. Environ. Res., 184, 109350. https://doi.org/10.1016/j.envres.2020.109350

Suresh, K, U Khanal, C Wilson et al. 2021. An economic analysis of agricultural adaptation to climate change impacts in Sri Lanka: An endogenous switching regression analysis. Land Use Policy, 109, 105601. https://doi.org/10.1016/j.landusepol.2021.105601

Sutton, WR, JP Srivastava, JE Neumann 2013. Looking beyond the horizon: how climate change impacts and adaptation responses will reshape agriculture in eastern Europe and Central Asia. The World Bank. https://doi.org/10.1596/978-0-8213-9768-8

Swanepoel, FJC, A Stroebel, S Moyo 2010. The Role of Livestock in Developing Communities: Enhancing Multifunctionality. University of the Free State and CTA. Available online: https://cgspace.cgiar.org/handle/10568/3003

Swapna P et al. 2020. Sea-Level Rise Assessment of Climate Change Over the Indian Region. Singapore: Springer Singapore, pp. 175–189.

Sweet, WV, BD Hamlington, RE Kopp et al. 2022. Global and Regional Sea Level Rise Scenarios for the United States: Updated Mean Projections and Extreme Water Level Probabilities Along U.S. Coastlines. NOAA Technical Report NOS 01. Silver Spring, MD: National Oceanic and Atmospheric Administration, National Ocean Service, 111 pp. https://www.usgs.gov/publications/global-and-regional-sea-level-rise-scenarios-united-states

Sweet, WV, R Horton, RE Kopp et al. 2017. Sea level rise. In: Wuebbles DJ, DW Fahey, KA Hibbard et al. (eds.), Climate Science Special Report: Fourth National Climate Assessment, Volume I. Washington, DC, USA: U.S. Global Change Research Program, pp. 333–363. https://doi.org/10.7930/J0VM49F2

Tabler, T, ML Khaitsa, J Wells et al. 2021. Poultry Production in Africa: Impacts of Climate Change. Mississippi State University Extension Service, Publication Number: P3706. Available at: https://extension.msstate.edu/publications/poultry-production-africa-impacts-climate-change

Tam, B, K Szeto, B Bonsal et al. 2018. CMIP5 drought projections in Canada based on the standardized precipitation evapotranspiration index. Can. Water Resour. J., 44, 90–107.

Tang, K, C He, C Ma et al. 2019. Does carbon farming provide a cost-effective option to mitigate GHG emissions? Evidence from China. Aust. J. Agric. Resour. Econ., 63, 575–592.

Tang, M, F Hu 2021. How does land urbanization promote CO_2 emissions reduction? Evidence from Chinese prefectural-level cities. Front. Environ. Sci., 606.

Tankson, JD, Y Vizzier-Thaxton, JP Thaxton et al. 2001. Stress and nutritional quality of broilers. Poult. Sci., 80, 1384–1389.

Tao, F, L Zhang, Z Zhang et al. 2022. Designing wheat cultivar adaptation to future climate change across China by coupling biophysical modelling and machine learning. Eur. J. Agron., 136, 126500.

Tao, F, M Yokozawa, Z Zhang 2009. Modelling the impacts of weather and climate variability on crop productivity over a large area: A new process-based model development, optimization, and uncertainties analysis. Agric. For. Meteorol., 149(5), 831–850. https://doi.org/10.1016/j.agrformet.2008.11.004

Tarr, AP, IJ Smith, CJ Rodger 2022. Carbon dioxide emissions from international air transport of people and freight: New Zealand as a case study. Environ. Res. Commun., 4, 075012.

Tate, E, MA Rahman, CT Emrich et al. 2021. Flood exposure and social vulnerability in the United States. Nat. Hazards, 106, 435–457. https://doi.org/10.1007/s11069-020-04470-2

Taylor, KE, RJ Stouffer, GA Meehl 2012. An overview of CMIP5 and the experiment design. Bull. Am. Meteorol. Soc., 93, 485–498.

Teng, PPS, M Caballero-Anthony, G Tian et al. 2015. Impact of Climate Change on Food Production: Options for Importing Countries. S. Rajaratnam School of International Studies. https://www.preventionweb.net/publication/impact-climate-change-food-production-options-importing-countries

Teng, PSS, M Caballero-Anthony, JA Lassa 2016. Projected Change in Rice Yield by 2030, 2050 and 2080. In The Future of Rice Security Under Climate Change (pp. 51–62). S. Rajaratnam School of International Studies. http://www.jstor.org/stable/resrep05930.11

Tesfaye, K, P Zaidi, S Gbegbelegbe et al. 2017. Climate change impacts and potential benefits of heat-tolerant maize in South Asia. Theor. Appl. Climatol., 130, 959–970.

Tessema, B, R Sommer, K Piikki et al. 2019. Potential for soil organic carbon sequestration in grasslands in East African countries: A review. Grassl. Sci., 1–10 https://doi.org/10.1111/grs.12267.

Thangataa, PH, PE Hildebrand 2012. Carbon stock and sequestration potential of agroforestry systems in smallholder agroecosystems of sub-Saharan Africa: Mechanisms for 'reducing emissions from deforestation and forest degradation' (REDD+). Agric. Ecosyst. Environ., 158, 172–183.

Thapa, G, R Gaiha 2014. Smallholder farming in Asia and the Pacific: challenges and opportunities, in: Hazell, PBR, A Rahman (Eds.). New Directions for Smallholder Agriculture. Oxford University Press, pp. 69–114. https://doi.org/10.1093/acprof:oso/9780199689347.003.0004

The World Bank 2022. World Bank Open Data. 2022. Available online: https://data.worldbank.org/

Theint, EE, S Suzuki, E Ono et al. 2015. Influence of different rates of gypsum application on methane emission from saline soil related with rice growth and rhizosphere exudation. Catena, 133, 467–473.

Thevathasan, NV, AM Gordon, R Bradley et al. 2012. Agroforestry research and development in Canada: The way forward. In: Nair P, D Garrity (eds) Agroforestry – The Future of Global Land Use. Advances in Agroforestry, vol. 9. Dordrecht: Springer. https://doi.org/10.1007/978-94-007-4676-3_15

Thiam, S, GB Villamor, LC Faye et al. 2021. Monitoring land use and soil salinity changes in coastal landscape: A case study from Senegal. Environ. Monit. Assess., 193, 259. https://doi.org/10.1007/s10661-021-08958-7

Thomas, TS, A Kamiljon, R Robertson et al. 2021. Climate change, agriculture, and potential crop yields in central Asia. IFPRI Discussion Paper 2081, Available at SSRN: https://ssrn.com/abstract=4000738

Thomas, TS, T Ponlok, R Bansok et al. 2013. Cambodian agriculture: Adaptation to climate change impact. Intl. Food Policy Res. Inst., 1285.

Thonicke, K, W Cramer 2006. Long-term trends in vegetation dynamics and forest fires in Brandenburg (Germany) under a changing climate. Nat. Hazards, 38, 283–300.

Thornton, PK 2010. Livestock production: Recent trends, future prospects. Philos. Trans. R. Soc. B, 365, 2853–2867.

Thornton, PK, M Herrero 2014. Climate change adaptation in mixed crop-livestock systems in developing countries. Glob. Food Secur., 3 (2), 99–107. https://doi.org/10.1016/j.gfs.2014.02.002

Thornton, PK, PG Jones, G Alagarswamy et al. 2010. Adapting to climate change: Agricultural system and household impacts in East Africa. Agric. Syst., 103(2), 73–82.

Tian, H, G Chen, C Lu et al. 2015. North American terrestrial CO_2 uptake largely offset by CH_4 and N_2O emissions: Toward a full accounting of the greenhouse gas budget. Clim. Change, 129, 413–426.

Timmerman, A, A Nygren, B VanDeWalle et al. 2023. History of Soybean Management. Cropwatch. University of Nebraska-Lincoln. https://cropwatch.unl.edu/soybean-management/history-soybean-management

Toensmeier, E 2016. The Carbon Farming Solution: a Global Toolkit of Perennial Crops and Regenerative Agriculture Practices for Climate Change Mitigation and Food Security. Chelsea Green Publishing.

Tol, RSJ 2021. Europe's climate target for 2050: An assessment. Int. Econ., 56(6), 330–335. https://doi.org/10.1007/s10272-021-1012-7

Tongwane, MI, ME Moeletsi 2018. A review of greenhouse gas emissions from the agriculture sector in Africa. Agric. Syst., 166, 124–134.

Tooley, BE, EB Mallory, GA Porter et al. 2021. Predicting the response of a potato-grain production system to climate change for a humid continental climate using DSSAT. Agric. For. Meteorol., 307, 108452.

Toulotte, JM, CK Pantazopoulou, MA Sanclemente et al. 2022. Water stress resilient cereal crops: Lessons from wild relatives. J. Integr. Plant Biol., 64(2), 412–430. https://doi.org/10.1111/jipb.13222

Tramberend, S, R Burtscher, P Burek et al. 2021. Co-development of East African regional water scenarios for 2050. One Earth, 4, 434–447. https://doi.org/10.1016/j.oneear.2021.02.012

Tramblay, Y, G Villarini, ME Saidi et al. 2022. Classification of flood-generating processes in Africa. Sci. Rep., 12, 18920. https://doi.org/10.1038/s41598-022-23725-5

Tramblay, Y, G Villarini, W Zhang 2020. Observed changes in flood hazard in Africa. Environ. Res. Lett., 15, 1040b5. https://doi.org/10.1088/1748-9326/abb90b

Tramblay, Y, S Somot 2018. Future evolution of extreme precipitation in the Mediterranean. Clim. Change, 151, 289–302. https://doi.org/10.1007/s10584-018-2300-5

Traore, B, K Descheemaeker, MT Van Wijk et al. 2017. Modelling cereal crops to assess future climate risk for family food self-sufficiency in southern Mali. Field Crop. Res., 201, 133–145.

Trenberth, KE 2011. Changes in precipitation with climate change. Clim. Res., 47(1), 123–138.

Trenberth, KE, A Dai, RM Rasmussen, DB Parsons 2003. The changing character of precipitation. Bull. Am. Meteorol. Soc., 84, 1205–1217.

Tubiello, F, M Salvatore, R Cóndor Golec et al. 2014. Agriculture, forestry and other land use emissions by sources and removals by sinks. Rome (Italy). https://www.uncclearn.org/wp-content/uploads/library/fao198.pdf.

Tubiello, FN, C Rosenzweig, RA Goldberg et al. 2002. Effects of climate change on US crop production: Simulation results using two different GCM scenarios. I. Wheat, potato, maize, and citrus. Clim. Res., 20, 259–270.

Tudge, SJ, A Purvis, A De Palma 2021. The impacts of biofuel crops on local biodiversity: A global synthesis. Biodivers. Conserv., 30, 2863–2883. https://doi.org/10.1007/s10531-021-02232-5

Tumeliené, E, J Sužiedelyté Visockiené, V Maliene 2022. Evaluating the eligibility of abandoned agricultural land for the development of wind energy in Lithuania. Sustainability, 14, 14569. https://doi.org/10.3390/su142114569

Twomlow, S, FT Mugabe, M Mwale et al. 2008. Building adaptive capacity to cope with increasing vulnerability due to climatic change in Africa – A new approach. Phys. Chem. Earth, 33, 780–787.

Tzanidakis, C, O Tzamaloukas, P Simitzis et al. 2023. Precision livestock farming applications (PLF) for grazing animals. Agriculture, 13, 288. https://doi.org/10.3390/agriculture13020288

UCAR 2023. Predictions of Future Global Climate. UCAR CENTER FOR SCIENCE EDUCATION. https://scied.ucar.edu/learning-zone/climate-change-impacts/predictions-future-global-climate

Uchida, Y, TJ Clough, FM Kelliher et al. 2011. Effects of bovine urine, plants and temperature on N_2O and CO_2 emissions from a sub-tropical soil. Plant Soil, 345, 171–186. https://doi.org/10.1007/s11104-011-0769-z

Udawatta, RP, S Jose 2012. Agroforestry strategies to sequester carbon in temperate North America. Agrofor. Syst., 86(2), 225–242.

Uhlenbrook, S, W Yu, P Schmitter et al. 2022. Optimising the water we eat—Rethinking policy to enhance productive and sustainable use of water in agri-food systems across scales. Lancet Planet. Health, 6(1), e59–e65.

Umirbekov, A, A Akhmetov, Z Gafurov 2022. Water-Agriculture-Energy Nexus in Central Asia Through the Lens of Climate Change. Central Asia Regional Economic Cooperation (CAREC) Institute.

UN DESA 2022. World Population Prospects 2022. https://population.un.org/wpp/. United Nations Department of Economic and Social Affairs.

Unc, A, D Altdorff, E Abakumov et al. 2021. Expansion of agriculture in northern cold-climate regions: A cross-sectoral perspective on opportunities and challenges. Front. Sustain. Food Syst., 5, 663448. https://doi.org/10.3389/fsufs.2021.663448

UNCCD 2015. Ghana national drought plan 1. United Nations Convention to Combat Desertification. https://knowledge.unccd.int/sites/default/files/country_profile_documents/1%2520FINAL_NDP_Ghana.pdf

UNCCD 2019. Land-based adaptation and resilience: powered by nature. Report retrieved from https://www.eld-initiative.org/fileadmin/pdf/Land_Based_Adaptation_ENG_Sall_web.pdf

UNCCD 2022. Drought in numbers 2022 – restoration for readiness and resilience. United Nations Convention to Combat Desertification. Available online at https://www.unccd.int/sites/default/files/2022-05/Drought_in_Numbers_%28English%29.pdf

UNCTAD 2021. Economic Development in Africa Report 2021: UNCTAD/PRESS/PR/2021/046. Available at: https://unctad.org/press-material/facts-and-figures-7

UNDRR 2021. United Nations Office for Disaster Risk Reduction 2021. GAR Special Report on Drought 2021. Geneva.

UNEP 2017. Global Environment Outlook 6 (GEO-6): Regional Assessment for West Asia. United Nations Environment Programme.

UNESCAP 2022. Opportunities for reaching carbon neutrality in the Asia-Pacific region. United Nations Economic and Social Council. https://www.unescap.org/sites/default/d8files/event-documents/ESCAP_78_11_E_.pdf

United Nations 2019. World population prospects: The 2019 revision. Department of Economic and Social Affairs, Population Division, United Nations. Online Edition. Rev. 1. Available at: https://population.un.org/wpp/Download/Standard/Population/.

United States Department of State and the United States Executive Office of the President 2021. The Long-Term Strategy of the United States: Pathways to Net-Zero Greenhouse Gas Emissions by 2050. Washington DC. November 2021. https://www.whitehouse.gov/wp-content/uploads/2021/10/US-Long-Term-Strategy.pdf

UN-Water 2013. Climate change (WWW document).

Unzilatirrizqi, Y, B Istiyanto, A Maulana 2019. The CO_2 Emissions Distribution Due to Contribution of Transportation Activities in Tegal City, Central Java. Advances in Engineering Research, 186, 11th Asia Pacific Transportation and the Environment Conference (APTE 2018).

US EPA 2016. U.S Environmental Protection Agency. MOVES2014a: Latest Version of Motor Vehicle Emission Simulator (MOVES) [WWW Document]. US EPA. URL https://www.epa.gov/moves/moves2014a-latest-version-motor-vehicle-emission-simulator-moves.

USAID 2017. Climate risk in Armenia: country risk profile. https://www.climatelinks.org/sites/default/files/asset/document/2017_USAID_Climate%20Change%20Risk%20Profile_Armenia.pdf

USDA 2011. USDA Agroforestry Strategic Framework, Fiscal Year 2011–2016. Washington, DC: U.S. Department of Agriculture.

USDA 2022. Agriculture and Food Statistics: Charting the Essentials. U.S. Department of Agriculture, Economic Research Service.

USDA-FAS (United States Department of Agriculture-Foreign Agricultural Service) 2015. Production, supply, and distribution online database. https://www.fas.usda.gov/data.

USDA-FAS 2021. Canada: grain and feed update. https://www.fas.usda.gov/data/canada-grain-and-feed-update-21.

USGCRP 2014. Shafer, M, D Ojima, JM Antle, D Kluck, RA McPherson, S Petersen, B Scanlon, K Sherman, 2014: Ch 19: Great Plains. Climate Change Impacts in the United States: The Third National Climate Assessment, J.M. Melillo, Terese (T.C.) Richmond, and G.W. Yohe, Eds., U.S. Global Change Research Program, pp. 441–461.

USGCRP 2018. Impacts, Risks, and Adaptation in the United States: Fourth National Climate Assessment, Volume II [D.R. Reidmiller, et al. (eds.)]. Washington, DC, USA: U.S. Global Change Research Program, 1515 pp. https://doi.org/10.7930/NCA4.2018

Vakulchuk, R, AS Daloz, I Overland I et al. 2022. A void in Central Asia research: Climate change. Cent. Asian Surv. https://doi.org/10.1080/02634937.2022.2059447

van der Velde, M, G Wriedt, F Bouraoui 2010. Estimating irrigation use and effects on maize yield during the 2003 heat wave in France. Agric. Ecosyst. Environ., 135, 90–97. https://doi.org/10.1016/j.agee.2009.08.017

van der Wiel, K, TJ Batelaan, N Wanders 2022. Large increases of multi-year droughts in north-western Europe in a warmer climate. Clim. Dyn. https://doi.org/10.1007/s00382-022-06373-3

van Dijk, M, T Morley, ML Rau et al. 2021. A meta-analysis of projected global food demand and population at risk of hunger for the period 2010–2050. Nat. Food, 2, 494–501. https://doi.org/10.1038/s43016-021-00322-9.

Van Eerd, LL, I Chahal, Y Peng et al. 2023. Influence of cover crops at the four spheres: A review of ecosystem services, potential barriers, and future directions for North America. Sci. Total Environ., 858(Pt 3), 159990. https://doi.org/10.1016/j.scitotenv.2022.159990

Van Grinsven, HJ, A Tiktak, CW Rougoor 2016. Evaluation of the Dutch implementation of the Nitrates Directive, the water framework Directive and the national emission ceilings Directive. NJAS – Wagening. J. Life Sci., 78, 69–84.

van Grinsven, HJM, JW Erisman, W de Vries et al. 2015. Potential of extensification of European agriculture for a more sustainable food system, focusing on nitrogen. Environ. Res. Lett., 10, 025002. https://doi.org/10.1088/1748-9326/10/2/025002

van Zanten, H, O van Hal, I de Boer 2016. The drivers of livestock production in the EU. SUSFANS DELIVERABLES. Deliverable No. 4.1.

Varma, V, DP Bebber 2019. Climate change impacts on banana yields around the world. Nat. Clim. Change. https://doi.org/10.1038/s41558-019-0559-9

Varshney, RK, R Barmukh, M Roorkiwal et al. 2021. Breeding custom-designed crops for improved drought adaptation. Adv. Genet., 2, e202100017.

Venkatramanan, V, S Shah, AK Rai et al. 2021. Nexus between crop residue burning, bio-economy and sustainable development goals over North-Western India. Front. Energy Res., 8, 392. https://doi.org/10.3389/fenrg.2020.614212

Vermeer, M, S Rahmstorf 2009. global sea level linked to global temperature. Proc. Natl. Acad. Sci. U. S. A., 106, 21527–21532.

Verschuuren, J 2022. Achieving agricultural greenhouse gas emission reductions in the EU post-2030: What options do we have? Rev. Eur. Comp. Int. Environ. Law (RECIEL), 31(2), 246–257.

Vincent, LA, X Zhang, E Mekis et al. 2018. Changes in Canada's climate: Trends in indices based on daily temperature and precipitation data. Atmos. – Ocean, 56(5), 332–349. https://doi.org/10.1080/07055900.2018.1514579

Vincent, LA, X Zhang, RD Brown et al. 2015. Observed trends in Canada's climate and influence of low-frequency variability modes. J. Clim., 28, 4545–4560. https://doi.org/10.1175/JCLI-D-14-00697.1

Vinke, K, MA Martin, S Adams et al. 2017. Climatic risks and impacts in South Asia: Extremes of water scarcity and excess. Reg. Environ. Change, 17, 1569–1583.

Vittersø, G, U Kjærnes, MH Austgulen 2015. Sustainable consumption in the Norwegian political economy of beef. In The Consumer in Society: A Tribute to Eivind Stø; Strandbakken, P., Gronow, J. (Eds.). Oslo, Norway: Abstrakt, pp. 267–290. ISBN 978-82-7935-369-0.

Vizy, EK, KH Cook 2012. Mid-twenty-first-century changes in extreme events over northern and tropical Africa. J Clim. https://doi.org/10.1175/JCLI-D-11-00693.1

Vogel, J 2022. Drivers of phenological changes in southern Europe. Int. J. Biometeorol., 66, 1903–1914 (2022). https://doi.org/10.1007/s00484-022-02331-0

von Braun, J, MS Sorondo, R Steiner 2023. Reduction of Food Loss and Waste: The Challenges and Conclusions for Actions. In: von Braun, J., Afsana, K., Fresco, L.O., Hassan, M.H.A. (eds) Science and Innovations for Food Systems Transformation. Cham: Springer. https://doi.org/10.1007/978-3-031-15703-5_31

Vose, RS, DR Easterling, KE Kunkel et al. 2017. Temperature changes in the United States. In: Climate Science Special Report: Fourth National Climate Assessment, Volume I [Wuebbles DJ, DW Fahey, KA Hibbard et al. (eds.)]. Washington, DC, USA:U.S. Global Change Research Program, pp. 185–206, https://doi.org/10.7930/J0N29V45

Vousdoukas, MI, L Mentaschi, E Voukouvalas et al. 2017. Extreme sea levels on the rise along Europe's coasts. Earths Future, 5, 304–323.

Vousdoukas, MI, L Mentaschi, E Voukouvalas et al. 2018. Feyen Climatic and socioeconomic controls of future coastal flood risk in Europe Nat. Clim. Change, 8 (9), 776–780, https://doi.org/10.1038/s41558-018-0260-4

Vousdoukas MI, L Mentaschi, J Hinkel et al. 2020. Economic motivation for raising coastal flood defenses in Europe Nat. Commun., 11 (1):1–11, https://doi.org/10.1038/s41467-020-15665-3

Vukicevich, E, T Lowery, P Bowen et al. 2016. Cover crops to increase soil microbial diversity and mitigate decline in perennial agriculture. A review. Agron. Sustain. Dev. 36, 48. https://doi.org/10.1007/s13593-016-0385-7.

Waha, K, MT Van Wijk, S Fritz et al. 2018. Agricultural diversification as an important strategy for achieving food security in Africa. Glob. Change Biol., 24, 3390–3400. https://doi.org/10.1111/gcb.14158

Waithaka, M, GC Nelson, TS Thomas et al. 2013. East African Agriculture and Climate Change: A Comprehensive Analysis. Washington, DC: International Food Policy Research Institute (IFPRI) https://doi.org/10.2499/9780896292055

Waldron, A, D Garrity, Y Malhi et al. 2017. Agroforestry can enhance food security while meeting other sustainable development goals. Trop. Conserv. Sci, 10, 1–6.

Waldron, A, DC Miller, D Redding et al. 2017. Reductions in global biodiversity loss predicted from conservation spending. Nature, 551, 364.

Wallach, D, D Makowski, J Jones 2006. Working With Dynamic Crop Models: Evaluation, Analysis, Parameterization, and Applications. Amsterdam, The Netherlands: Elsevier.

Wallander, S 2020. While Crop Rotations Are Common, Cover Crops Remain Rare. USDA 2020. https://www.ers.usda.gov/amber-waves/2013/march/while-crop-rotations-are-common-cover-crops-remain-rare/

Wang, F, JD Harindintwall, ZZ Yuan et al. 2021. Technologies and perspectives for achieving carbon neutrality. Innovation, 2(4), 100180.

Wang, H, C Yue, Q Mao et al. 2020. Vegetation and species impacts on soil organic carbon sequestration following ecological restoration over the Loess Plateau. China Geoderma, 371, 114389.

Wang, J, J Huang, J Yang 2014. Overview of impacts of climate change and adaptation in China's agriculture. J. Integr. Agric., 13, 1–17.

Wang, J, Z Xiong, Y Kuzyakov 2016. Biochar stability in soil: Meta-analysis of decomposition and priming effects Glob. Change Biol. Bioenergy, 8, 512–523.

Wang, M, Y Wu, A Elgowainy 2007. Operating Manual for GREET: Version 1.7 (Research report No. ANL/ESD/05-3).

Wang, R, LC Bowling, KA Cherkauer et al. 2017. Biophysical and hydrological effects of future climate change including trends in CO_2, in the St. Joseph River watershed, Eastern Corn Belt. Agric. Water Manage., 180, 280–296. https://doi.org/10.1016/j.agwat.2016.09.017.

Wang, SL, P Heisey, D Schimmelpfennig et al. 2015. Agricultural Productivity Growth in the United States: Measurement, Trends, and Drivers, ERR-189, U.S. Department of Agriculture, Economic Research Service, July 2015.

Wang, WG, GG Zheng 2012. Annual Report on Actions to Address Climate Change: Climate Finance and Low Carbon Development. Social Science Academic Press (in Chinese).

Wang, Z, Y Rasool, MM Asghar et al. 2019. Dynamic linkages among CO_2 emissions, human development, financial development, and globalization: Empirical evidence based on PMG long-run panel estimation. Environ. Sci. Pollut. Res. https://doi.org/10.1007/s11356-019-06556-2

Ward, CR, DR Chadwick, PW Hill 2023. Potential to improve nitrogen use efficiency (NUE) by use of perennial mobile green manures. Nutr. Cycl. Agroecosyst., 125, 43–62. https://doi.org/10.1007/s10705-022-10253-x

Ward, P, F Raymond, A Flores-Lagunes (2011). Climate change and agricultural productivity in sub-Saharan Africa: a spatial sample selection model, Dept. of Agricultural Economics. Purdue University Working Paper: 11-4.

Ward, PS, RJGM Florax, A Flores-Lagunes 2011. Climate Change and Agricultural Productivity in Sub-Saharan African: a Spatial Sample Selection Model. Working Paper # 11–4. Dept. of Agricultural Economics at Purdue University.

Wardah, B, Toknok, Zulkhaidah 2013. Carbon stock of agroforestry systems at adjacent buffer zone of Lore Lindu National Park, Central Sulawesi. J. Trop. Soils, 16, 123–128.

Warren, FJ, DS Lemmen 2014. Synthesis—Canada in a changing climate: sector perspectives on impacts and adaptation, Government of Canada, Ottawa, ON, pp. 1–18.

WCRP Global Sea Level Budget Group 2018. Global sea-level budget 1993–present. Earth Syst. Sci. Data, 10, 1551–1590. https://doi.org/10.5194/essd-10-1551-2018

Webber, H, F Ewert, JE Olesen et al. 2018. Diverging importance of drought stress for maize and winter wheat in Europe. Nat. Commun., 9(1), 4249.

Webber, H, G Lischeid, M Sommer et al. 2020. No perfect storm for crop yield failure in Germany. Environ. Res. Lett., 15(104012). https://doi.org/10.1088/1748-9326/aba2a4

Wehner, MF, JR Arnold, T Knutson et al. 2017. Droughts, Floods, and Wildfires. Climate Science Special Report: Fourth National Climate Assessment vol. I, D J Wuebbles, D W Fahey, K A Hibbard, D J Dokken, B C Stewart and T K Maycock (Eds.). Washington, DC: US Global Change Research Program, pp 231–56. https://doi.org/10.7930/J0CJ8BNN

Weih, M, E Adam, G Vico et al. 2022. Application of crop growth models to assist breeding for intercropping: Opportunities and challenges. Front. Plant Sci., 13, 720486. https://doi.org/10.3389/fpls.2022.720486

Weisse, R, D Bellafiore, M Menéndez et al. 2014. Willems Changing extreme sea levels along European coasts. Coast. Eng., 87; 4–14, https://doi.org/10.1016/j.coastaleng.2013.10.017.

Weissman, DS, KL Tully 2020. Saltwater intrusion affects nutrient concentrations in soil porewater and surface waters of coastal habitats. Ecosphere, 11 (2), e03041. https://doi.org/10.1002/ECS2.3041

Welegedara, NP, SK Agrawal, S Gajjar et al. 2021. Variations in direct greenhouse gas emissions across neighbourhoods: A case of Edmonton in Canada. Environ. Chall., 5, 100312. https://doi.org/10.1016/j.envc.2021.100312

Wentz, FJ, L Ricciardulli, K Hilburn et al. 2007. How much more rain will global warming bring? Science, 317(5835), 233–235. https://doi.org/10.1126/science.1140746

WHO 2020. State of the Climate in Africa 2019. World Meteorological Organization, 2020. Report N° WMO-No. 1253. https://library.wmo.int/doc_num.php?explnum_id=10421

Wiebe, K, H Lotze-Campen, R Sands et al. 2015. Climate change impacts on agriculture in 2050 under a range of plausible socioeconomic and emissions scenarios. Environ. Res. Lett., 10, 085010.

Wilby, RL, CW Dawson, EM Barrow 2002. SDSM—A decision support tool for the assessment of regional climate change impacts. Environ. Modell. Softw, 17, 145–157.

Wilson, M, S Lovell 2016. Agroforestry-the next step in sustainable and resilient agriculture. Sustainability, 8(6), 574.

Wing, IS, E De Cian, MN Mistry 2021. Global vulnerability of crop yields to climate change. J. Environ. Econ. Manag., 109, 102462.

Winiwarter, W, J Hettelingh, L Bouwman et al. 2011. Future Scenarios of Nitrogen in Europe. In: Sutton M., Howard C., Erisman J. W., et al. 2011. European Nitrogen Assessment. Cambridge (United Kingdom): Cambridge University Press; 2011. pp. 551–569. JRC65635.

Wiréhn, L 2018. Nordic agriculture under climate change: A systematic review of challenges, opportunities and adaptation strategies for crop production. Land Use Policy 77, 63–74. https://doi.org/10.1016/j.landusepol.2018.04.059

WMO 2016. Statement on the status of the global climate in 2015. World Meteorological Organization. Retrieved from http://public.wmo.int/en/media/press-release/provisional-wmo-statement-status-of-global-climate-2016

World Bank 2010. The Economics of Adaptation to Climate Change (EACC): Synthesis Report. Washington, DC: The World Bank Group.

World Bank 2020. Climate Change Knowledge Portal for Development Practitioners and Policy Makers. https://climateknowledgeportal.worldbank.org/.

World Bank 2022. Pakistan: Flood Damages and Economic Losses Over USD 30 billion and Reconstruction Needs Over USD 16 billion – New Assessment. PRESS RELEASE NO: SAR/2022.

World Health Organization 2014. Quantitative Risk Assessment of the Effects of Climate Change on Selected Causes of Death, 2030s and 2050s. World Health Organization. https://apps.who.int/iris/handle/10665/134014.

Wray-Cahen, D, A Bodnar, C Rexroad et al. 2022. Advancing genome editing to improve the sustainability and resiliency of animal agriculture. CABI Agric. Biosci., 3, 21. https://doi.org/10.1186/s43170-022-00091-w

Wriedt, G, M van der Velde, A Aloe et al. 2009. European irrigation map for spatially distributed agricultural modelling. Agric. Water Manag., 96: 771–789, https://doi.org/10.1016/j.agwat.2008.10.012

Wright, I, S Tarawali, M Blummel et al. 2012. Integrating crops and livestock in subtropical agricultural systems. J. Sci. Food Agric., 92, 1010–1015.

Wu, L, A Elshorbagy, W Helgason 2023. Assessment of agricultural adaptations to climate change from a water-energy-food nexus perspective. Agric. Water Manag., 284, 108343. https://doi.org/10.1016/j.agwat.2023.108343

Wynants, M, C Kelly, K Mtei et al. 2019. Drivers of increased soil erosion in East Africa's agro-pastoral systems: Changing interactions between the social, economic and natural domains. Reg. Environ. Change, 19, 1909–1921.

Xie, W, J Huang, J Wang et al. 2020. Climate change impacts on China's agriculture: The responses from market and trade. China Econ. Rev., 62, 101256.

Xiong, J, PS Thenkabail, MK Gumma et al. 2017. Automated cropland mapping of continental Africa using Google Earth Engine cloud computing. ISPRS J. Photogramm. Remote Sens., 126, 225–244.

Xu, F, B Wang, C He et al. 2021. Optimizing sowing date and planting density can mitigate the impacts of future climate on maize yield: A case study in the Guanzhong Plain of China. Agronomy, 11, 1452. https://doi.org/10.3390/agronomy11081452

Xu, Y, X Gao, Y Shi et al. 2015. Detection and attribution analysis of annual mean tempera-ture changes in China. Clim. Res., 63, 61–71.

Xue, Q, TH Marek, W Xu et al. 2017. Irrigated corn production and management in the Texas high plains. J. Contemp. Water Res. Educ., 162(1), 31–41.

Yan, S, S Alvi 2022. Food security in South Asia under climate change and economic poli-cies. Int. J. Clim. Change Strateg. Manag., 14 (3), 237–251. https://doi.org/10.1108/IJCCSM-10-2021-0113

Yang, C, H Fraga, W van Ieperen et al. 2019a. Effects of climate change and adaptation options on winter wheat yield under rainfed Mediterranean conditions in southern Portugal. Clim. Change, 154, 159–178. https://doi.org/10.1007/s10584-019-02419-4

Yang, F, Z Liu, Y Wang et al. 2023a A variety test platform for the standardization and data quality improvement of crop variety tests. Front. Plant Sci., 14, 1077196. https://doi.org/10.3389/fpls.2023.1077196

Yang, JM, JY Yang, S Liu et al. 2014. An evaluation of the statistical methods for testing the performance of crop models with observed data. Agric. Syst., 127, 81–89.

Yang M, L Chen, J Wang et al. 2023b Circular economy strategies for combating climate change and other environmental issues. Environ. Chem. Lett., 21, 55–80. https://doi.org/10.1007/s10311-022-01499-6

Yang Y, D Tilman, G Furey et al. 2019b Soil carbon sequestration accelerated by res-toration of grassland biodiversity. Nat. Commun., 10, 718. https://doi.org/10.1038/s41467-019-08636-w.

Yang, Y, SE Hobbie, RR Hernandez et al. 2020. Restoring abandoned farmland to mitigate climate change on a full earth. One Earth, 3, 176–186.

Yang YU, PI Yuanyue, YU Xiang et al. 2019c Climate change, water resources and sustain-able development in the arid and semi-arid lands of Central Asia in the past 30 years. J. Arid Land, 11(1), 1–14. https://doi.org/10.1007/s40333-018-0073-3

Yao, T, T Bolch, D Chen et al. The imbalance of the Asian water tower. Nat. Rev. Earth Environ., 3, 618–632 (2022). https://doi.org/10.1038/s43017-022-00299-4

Yatagai, A, K Kamiguchi, O Arakawa, A et al. 2012. APHRODITE: Constructing a long-term daily gridded precipitation dataset for Asia based on a dense network of rain gauges Bull. Am. Meteorol. Soc., 93 (9), 1401–1415, https://doi.org/10.1175/bams-d-11-00122.1

Ye, L, K Shi, Z Xin et al. 2019. Compound droughts and heat waves in China. Sustainability, 11, 3270. https://doi.org/10.3390/su11123270

Ye, L, W Xiong, Z Li et al. 2013. Climate change impact on China food security in 2050. Agron. Sustain. Dev., 33, 363–374 https://doi.org/10.1007/s13593-012-0102-0

Yerlikaya, BA, S Ömezli, N Aydoğan 2020. Climate change forecasting and modeling for the year of 2050. In Environment, Climate, Plant and Vegetation Growth (pp. 109–122). Cham: Springer. https://doi.org/10.1007/978-3-030-49732-3_5

Yoruk, E, EN Keles, O Sefer et al. 2018. Salinity and drought stress on barley and wheat cultivars planted in Turkey. J. Environ. Biol., 39, 943–950.

You, Q, Z Jiang, X Yue et al. 2022. Recent frontiers of climate changes in East Asia at global warming of 1.5°C and 2°C. NPJ Clim. Atmos. Sci., 5, 80.

Yu, C, R Miao, M Khanna 2021. Maladaptation of U.S. corn and soybeans to a changing climate. Sci. Rep., 11, 12351. https://doi.org/10.1038/s41598-021-91192-5.

Yu L, W Wang, L Niu et al. 2018. A New Cultivation Technique of Cangmai 6005 for High Yield in Cangzhou Dry-Alkali Land. Asian Agricultural Research: USA-China Science and Culture Media Corporation. Available from: https://ideas.repec.org/a/ags/asagre/273103.html

Yu, P, X Li, PJ White et al. 2015. A large and deep root system underlies high nitrogen-use efficiency in maize production. PLOS One, 10, e0126293. https://doi.org/10.1371/journal.pone.0126293

Yu S, Z Yan, N Frechet et al. 2020. Trends in summer heatwaves in central Asia from 1917 to 2016: Association with large-scale atmospheric circulation patterns. Int. J. Climatol., 40, 115–127, https://doi.org/10.1002/joc.6197

Yuan S, AM Stuart, AG Laborte et al. 2022. Southeast Asia must narrow down the yield gap to continue to be a major rice bowl. Nat. Food, 3, 217–226. https://doi.org/10.1038/s43016-022-00477-z

Yue, Q, X Xu, J Hillier, K Cheng, G Pan. 2017. Mitigating greenhouse gas emissions in agriculture: From farm production to food consumption. J. Clean. Prod., 149: 1011–1019. https://doi.org/10.1016/j.jclepro.2017.02.172

Zajac Z, O Gomez, E Gelati et al. 2022. Estimation of spatial distribution of irrigated crop areas in Europe for large-scale modelling applications. Agric. Water Manag., 266, 107527. https://doi.org/10.1016/j.agwat.2022.107527

Zerriffi, H, R Reyes, A Maloney 2023. Pathways to sustainable land use and food systems in Canada. Sustain. Sci., 18, 389–406. https://doi.org/10.1007/s11625-022-01213-z

Zhai, F, J Zhuang 2009. Agricultural Impact of Climate Change: A General Equilibrium Analysis with Special Reference to Southeast Asia. ADBI Working Paper 131. Tokyo: Asian Development Bank Institute. http://www.adbi.org/workingpaper/2009/02/23/2887.agricultural.impact.climate.change/

Zhang, L, H Tian, H Shi et al. 2021. Methane emissions from livestock in East Asia during 1961–2019. Ecosyst. Health Sustain., 7(1), 1918024. https://doi.org/10.1080/20964129.2021.1918024

Zhang, L, Y Xu, C Meng et al. 2020. Comparison of statistical and dynamic downscaling techniques in generating high-resolution temperatures in China from CMIP5 GCMs. J. Appl. Meteorol. Climatol., 59(2), 207–235.

Zhang M, Y Chen, Y Shen et al. 2019. Tracking climate change in Central Asia through temperature and precipitation extremes. J. Geogr. Sci., 29, 3–28. https://doi.org/10.1007/s11442-019-1581-6

Zhang, PF, GY Ren, Y Xu et al. 2019. Observed change in extreme temperature over the global land based on the newly developed CMA daily dataset. J. Clim., 32, 8489–8509. https://doi.org/10.1175/JCLI-D-18-0733.1

Zhang, R, B Wu, J Han et al. 2013. Effects on summer monsoon and rainfall change over China duo to Eurasian snow cover and ocean thermal conditions. Clim. Change Realities, Impacts Over Ice Cap, Sea Level Risks. https://doi.org/10.5772/54831

Zhang, X, EA Davidson, DL Mauzerall et al. 2015. Managing nitrogen for sustainable development. Nature, 528, 51. https://doi.org/10.1038/nature15743

Zhang X, G Flato, M Kirchmeier-Young et al. 2019. Changes in Temperature and Precipitation Across Canada; Chapter 4 in Bush, E. and Lemmen, D.S. (Eds.) Canada's Changing Climate Report. Government of Canada, Ottawa, Ontario, pp 112–193.

Zhang X, S Zhou, J Bi et al. 2021. Drought-resistance rice variety with water-saving management reduces greenhouse gas emissions from paddies while maintaining rice yields. Agric. Ecosyst. Environ., 320, 107592. https://doi.org/10.1016/j.agee.2021.107592

Zhao, C, B Liu, S Piao et al. 2017. Temperature increase reduces global yields of major crops in four independent estimates, Proc. Natl. Acad. Sci. U. S. A., 114, 9326–9331, https://doi.org/10.1073/pnas.1701762114.

Zhao, J, D Liu, R Huang 2023. A review of climate-smart agriculture: Recent advancements, challenges, and future directions. Sustainability, 15, 3404. https://doi.org/10.3390/su15043404

Zhao, J, M Bindi, J Eitzinger et al. 2022. Priority for climate adaptation measures in European crop production systems. Eur. J. Agron., 138, 126516.

Zhao, T, A Dai 2015. The magnitude and causes of global drought changes in the twenty-first century under a low–moderate emissions scenario. J. Clim., 28(11), 4490–4512. https://doi.org/10.1175/JCLI-D-14-00363.1

Zhao, X, S Pan, Z Liu et al. 2022. Intelligent upgrading of plant breeding: Decision support tools in the golden seed breeding cloud platform. Comput. Electron. Agric., 194, 106672. https://doi.org/10.1016/j.compag.2021.106672

Zhou, Q, DJ Soldat 2022. Evaluating decision support tools for precision nitrogen management on creeping bentgrass putting greens. Front. Plant Sci., 13, 863211. https://doi.org/10.3389/fpls.2022.863211

Zhou, YQ, GY Ren 2011. Change in extreme temperature events frequency over mainland China during 1961–2008. Clim. Res., 50(1–2), 125–139. https://doi.org/10.3354/cr01053

Ziervogel G, A Cartwright, A Tas et al. 2008. Climate Change and Adaptation in African Agriculture. Prepared for Rockefeller Foundation. Stockholm Environment Institute.

Zilberman, D, L Lipper, N McCarthy et al. 2018. Innovation in Response to Climate Change. In: Lipper L, N McCarthy, D Zilberman et al. (eds.), Climate Smart Agriculture. Natural Resource Management and Policy, vol 52. Cham: Springer. https://doi.org/10.1007/978-3-319-61194-5_4

Zinyengere, N, O Crespo, S Hachigonta 2013. Crop response to climate change in southern Africa: A comprehensive review. Glob. Planet Change, 111, 118–126. https://doi.org/10.1016/j.gloplacha.2013.08.010

Zittis, G, A Bruggeman, J Lelieveld 2021. Revisiting future extreme precipitation trends in the Mediterranean. Weather Clim. Extrem., 34, 100380.

Index

Printed in the United States
by Baker & Taylor Publisher Services